First Siberian Winter School
Algebra and Analysis

American Mathematical Society
TRANSLATIONS
Series 2 • Volume 148

Soviet Regional Conferences

First Siberian Winter School
Algebra and Analysis

Proceedings of the First Siberian School
Kemerovo State University, Kemerovo
1988

Editors
A. D. Aleksandrov
O. V. Belegradek, *Executive Editor*
L. A. Bokut'
Yu. L. Ershov

American Mathematical Society
Providence, Rhode Island

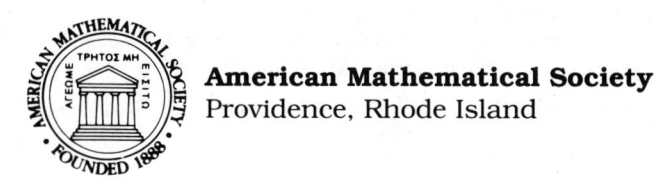

Translation edited by SIMEON IVANOV

1980 *Mathematics Subject Classification* (1985 *Revision*). Primary 05B35, 14L30, 15A72, 16A27, 16A38, 17B10, 17A70, 17B70, 22D40, 22E45, 22E46, 28D05, 34C25, 51M10, 51M20, 51M25, 53C65; Secondary 03C45, 03C60, 05B25, 11Q15, 14A25, 16A03, 16A46, 17A65, 20G05, 32L10, 52A55, 57Q45, 58A05, 58C99, 58F21, 81E13, 83E99.

Library of Congress Cataloging-in-Publication Data
Siberian Winter School "Algebra and Analysis" (1st: 1987: Kemerovo, R.S.F.S.R.).
   [Algebra i analiz. English]
   First Siberian Winter School "Algebra and Analysis": proceedings of the First Siberian School, Kemerovo State University, Kemerovo.
   p. cm. – (American Mathematical Society translations; ser. 2, v. 148. Soviet regional conferences)
   Translation of: Algebra i analiz.
   Includes bibliographical references.
   ISBN 0-8218-3700-1
   1. Algebra–Congresses. 2. Mathematical analysis–Congresses. I. Title. II. Series: American Mathematical Society translations; ser. 2, v. 148. III. Series: American Mathematical Society translations. Soviet regional conferences.
QA3.A572 ser. 2, vol. 148
[QA150]
510 s–dc20   90-23506
[512]   CIP
                                                                                   r91

    COPYING AND REPRINTING. Individual readers of this publication, and nonprofit libraries acting for them, are permitted to make fair use of the material, such as to copy an article for use in teaching or research. Permission is granted to quote brief passages from this publication in reviews, provided the customary acknowledgment of the source is given.
    Republication, systematic copying, or multiple reproduction of any material in this publication (including abstracts) is permitted only under license from the American Mathematical Society. Requests for such permission should be addressed to the Manager of Editorial Services, American Mathematical Society, P.O. Box 6248, Providence, Rhode Island 02940-6248.
    The appearance of the code on the first page of an article in this book indicates the copyright owner's consent for copying beyond that permitted by Sections 107 or 108 of the U.S. Copyright Law, provided that the fee of $1.00 plus $.25 per page for each copy be paid directly to the Copyright Clearance Center, Inc., 27 Congress Street, Salem, Massachusetts 01970. This consent does not extend to other kinds of copying, such as copying for general distribution, for advertising or promotional purposes, for creating new collective works, or for resale.

Copyright © 1991 by the American Mathematical Society. All rights reserved.
Printed in the United States of America.
The American Mathematical Society retains all rights
except those granted to the United States Government.
The paper used in this book is acid-free and falls within the guidelines
established to ensure permanence and durability. ⊚
This publication was typeset using $\mathcal{A}_{\mathcal{M}}\mathcal{S}$-TEX,
the American Mathematical Society's TEX macro system.

10 9 8 7 6 5 4 3 2 1   96 95 94 93 92 91

Contents

Preface	vii
A. M. Vershik, Local Stationary Algebras	1
È. B. Vinberg, The Volume of Polyhedra on a Sphere and in Lobachevsky Space	15
S. G. Gindikin, Integral Geometry on Symmetric Manifolds	29
E. I. Zel'manov, Superalgebras and Identities	39
B. I. Zil'ber, Algebraic Geometry and Combinatorics: a Model-Theoretic Point of View	47
Yu. S. Il'yashenko, Finiteness Theorems for Limit Cycles	55
A. R. Kemer, Identities of Associative Algebras	65
G. A. Margulis, Lie Groups and Ergodic Theory	73
A. L. Onishchik, Lie Superalgebras of Vector Fields	87
V. L. Popov, Invariant Theory	99

Preface

This collection consists of lectures delivered at the *First Siberian Winter School "Algebra and Analysis"* convened 9–17 March 1987 at a retreat not far from Kemerovo.

The school was organized by the Kemerovo State University and the Institute of Mathematics of the Siberian Branch of the Academy of Sciences of the USSR. More than 100 mathematicians from Novosibirsk, Kemerovo, Omsk, Moscow, Leningrad and other cities of the country participated.

The aim of the school was to organize a system of contacts between algebraists and specialists working in the area of mathematical analysis, in the broad sense of this word. The lectures acquainted the participants of the school with current research in algebra and areas in the boundary between algebra and analysis.

EDITORIAL BOARD:
A. D. Aleksandrov
O. V. Belegradek (Executive Editor)
L. A. Bokut'
Yu. L. Ershov

UDC 515.162.8

Local Stationary Algebras

A. M. VERSHIK

These notes contain an expanded part of a lecture delivered at the First Siberian School in Algebra and Analysis (Kemerovo, 1987). The author regarded the school as an attempt to bring together algebraists and mathematicians interested in analysis in the broad sense of the word and to give some current examples of how they could assist each other in solving new problems. One such example is undoubtedly the representation theory of locally semisimple algebras—a theory with purely algebraic formulations, but requiring the use of analytic and combinatorial methods. Proof of its relevance is, in particular, the fact that recent progress in knot theory and parallel investigations of certain classes of solutions of the Yang-Baxter equation have been based on the representation theory of Hecke algebras and, in particular, of the group algebra of the infinite symmetric group. We cannot indicate here the full scope of the subject[1], hence we choose a fragment which, in the author's opinion, is now of the greatest interest: local stationary algebras and their local representations (the definitions are given below). Among the topics related to this subject are the representation theory of infinite-dimensional classical groups and some problems of quantum field theory.

1. Local stationary algebras. Examples

Suppose an algebra A with unity over a field k is defined by generators $\sigma_1, \sigma_2, \ldots$ and relations. It is called local (it would be more precise to call it an algebra with "local interaction of generators") if $\sigma_i \sigma_{i+k} = \sigma_{i+k} \sigma_i$ for $k > k_0$, $i \geq 1$. We will be interested in the case $k_0 = 1$, i.e. the case where the generators "interact" only with their nearest neighbors. In fact this

---

1980 *Mathematics Subject Classification* (1985 Revision). Primary 16A27; Secondary 20G05, 57Q45, 81E13.

[1]We refer those interested to the surveys [1], [2], the papers of Jones [3], [4], and the literature cited therein.

assumption does not diminish the generality. An algebra is called stationary if the relations connecting σ_i and σ_{i+1} (if there are any) are the same for all i, i.e. for each i the mapping $\sigma_1 \mapsto \sigma_i$, $\sigma_2 \mapsto \sigma_{i+1}$ can be extended to an algebra isomorphism of $\mathrm{Alg}(\sigma_1, \sigma_2)$ and $\mathrm{Alg}(\sigma_i, \sigma_{i+1})$. This definition suffices for what follows, but we also mention the more general invariant concepts. Suppose there are given in an algebra A a subalgebra A_1 and an epimorphism $T: A_1 \to A$. If $\bigcap_n T^{-n} A_1 = \{\mathscr{k}\}$, then the algebra is called stationary. If there exists a subalgebra $B \subset A$, $A_1 \cap B = \{\mathscr{k}\}$, $A = \mathrm{Alg}\{A_1, B\}$, such that B and $T^{-2}A_1$ commute (and therefore $T^{-1}B$, $T^{-2}B$, ..., $T^{-k}B$ commute with $T^{-k-2}A$), then the algebra A with distinguished subalgebra B and endomorphism T is called a local stationary algebra. In the previous definition,

$$A = \mathrm{Alg}\{\sigma_1, \sigma_2, \ldots\}, \qquad A_1 = \mathrm{Alg}\{\sigma_2, \sigma_3, \ldots\},$$
$$B = \mathrm{Alg}\{\sigma_1\}, \qquad T\sigma_i = \sigma_{i-1}, \quad i \geq 2.$$

Of special interest are those local stationary algebras for which $\mathrm{Alg}\{\sigma_1, \ldots, \sigma_n\}$ is finite-dimensional and semisimple for all n. In this case we are in the category of locally semisimple algebras, which have been intensively studied in recent years (see [1] and its bibliography).

EXAMPLES. 1. Suppose for $k \geq 2$, $i \geq 1$ that $\sigma_i \sigma_{i+k} = \sigma_{i+k} \sigma_i$ and

$$\sigma_i \sigma_{i+1} \sigma_i = \sigma_{i+1} \sigma_i \sigma_{i+1}. \tag{$*$}$$

Then $\mathrm{Alg}\{\sigma_1, \ldots, \sigma_n\}$ is the group algebra $\mathscr{k}(B_n)$ of the group B_n, and $\mathrm{Alg}(\sigma_1, \sigma_2, \ldots)$ is the group algebra $\mathscr{k}(B_\infty)$ of the infinite (stable) group B_∞.

2. If in addition to the above relations we have $\sigma_i^2 = e$, we obtain the group algebra of the infinite symmetric group S_∞, i.e. the group of finite permutations of the natural sequence. In this case the relation $(*)$ can be written in the form $(\sigma_i \sigma_{i+1})^3 = e$.

3. If we introduce the relations $\sigma_i^2 + (1-q)\sigma_i - q = 0$, then $\mathrm{Alg}(\sigma_1, \ldots, \sigma_n\}$ is the classical Hecke algebra $H_n(q)$, and $\mathrm{Alg}(\sigma_1, \sigma_2, \ldots\} = H_\infty(q)$ is the infinite Hecke algebra. When $q = 1$ we obtain Example 2, and when $q = p^k$ we have the Hecke algebra in the sense of the theory of algebraic groups [5], [6].

4. Consider for $k \geq 2$, $i \geq 1$ the relations

$$e_i^2 = e_i, \quad e_i e_{i+1} e_i - e_{i+1} e_i e_{i+1} = \tau(e_i - e_{i+1}), \quad e_i e_{i+k} = e_{i+k} e_i.$$

The corresponding algebra is a Hecke algebra, which can be seen by putting for $i \geq 1$

$$e_i = \frac{\sigma_i + 1}{q + 1}, \qquad \tau = (q + q^{-1})^{-2}.$$

In contrast to $\mathscr{k}(B_n)$, we have that $\mathrm{Alg}\{e_1, \ldots, e_n\}$ is finite-dimensional and semisimple, and $\mathrm{Alg}\{e_1, e_2, \ldots\}$ is locally semisimple. A quotient algebra of the latter, sometimes called the Jones algebra, is obtained by considering for $k \geq 2$, $i \geq 1$ the relations

$$e_i^2 = e_i, \quad e_i e_{i+1} e_i = \tau e_i, \quad e_{i+1} e_i e_{i+1} = \tau e_{i+1}, \quad e_i e_{i+k} = e_{i+k} e_i.$$

Both of these algebras and their representations play a crucial role in recent investigations in knot theory. Of course, they are quotient algebras of $\mathscr{k}(S_\infty)$.

5. Suppose $\sigma_i^2 = e$ and there are no other relations. Then $\mathrm{Alg}\{\sigma_1, \sigma_2\}$ is the group algebra of the free product $\mathbb{Z}_2 \times \mathbb{Z}_2$, and $\mathrm{Alg}\{\sigma_1, \ldots, \sigma_n\}$ and $\mathrm{Alg}\{\sigma_1, \sigma_2, \ldots\}$ are very interesting algebras, apparently not studied in the literature:

•——∞——•——∞——• · · ·

6. Suppose $\sigma_i^2 = e$ and $(\sigma_i \sigma_{i+1})^k = e$. These relations define a quotient algebra of the preceding algebra. It is known from the theory of Coxeter groups [5] that $\mathrm{Alg}\{\sigma_1, \ldots, \sigma_n\}$ is finite-dimensional only for $k \leq 3$; when $k \geq 4$ the corresponding groups are infinite (if the number n of generators is at least 4).

The following example is particularly interesting.

7. Suppose for $k \geq 2$, $i \geq 1$ that

$$\sigma_i^2 = e, \quad P(\sigma_i \sigma_{i+1}) = 0, \quad \sigma_i \sigma_{i+k} = \sigma_{i+k} \sigma_i,$$

where $P(z)$ is a polynomial over $\mathscr{k}$.

For example, if

$$P(z) = (z-1)(z^2 + 2(1-2\tau)z + 1), \quad \tau = (q + q^{-1})^{-2},$$

then the corresponding algebra is again a Hecke algebra. A substitution—the new generators can be expressed in terms of the old (see Example 3) as follows:

$$\bar{\sigma}_k := \frac{2}{q+1}(\sigma_k + 1) - 1.$$

CONJECTURE. $\mathrm{Alg}\{\sigma_1, \ldots, \sigma_k\}$ in this example is finite-dimensional if and only if the polynomial P has degree at most 3.

È. B. Vinberg has argued that the "if" part can be proved in the same way as for Coxeter groups (i.e. for the case $P(z) = z^3 - 1$).

2. Representations

The main problem can be formulated as follows: find a sufficiently ample supply of natural modules over local stationary algebras, find formulas for the characters, etc. We will elaborate later on what is meant by "natural" in this case. From the standpoint of mathematical physics the problem consists of finding solutions of commutational relations of Yang-Baxter type. We will use the terminology of representation theory.

Our constructions will be inductive, i.e. representations (modules) over $\text{Alg}\{\sigma_1, \ldots, \sigma_n\}$ will be constructed as sums of modules over $\text{Alg}\{\sigma_1, \ldots, \sigma_{n-1}\}$. Here it is useful to have the concept of a Gel'fand-Tsetlin basis, which has a meaning in any locally semisimple algebra (see [1]). Suppose $k = A_0 \subset A_1 \subset \ldots$ is an ascending chain of semisimple algebras and M is a module over A_n. Viewing M as an A_{n-1}-module, we consider its constituent simple modules $M_{n-1}^1, \ldots, M_{n-1}^r$. If a basis $\{e_j^i\}$ of each module M_{n-1}^i has been chosen, we construct a new basis as follows: we write

$$M = \sum_i M_{n-1}^i \otimes k^{s_i},$$

where s_i is the multiplicity of the occurrence of M_{n-1}^i in M^i, and take any basis $\{f_t\}$ of k^{s_i}. The choice of basis is unique to within the group $GL(s_i, k)$. Thus we have constructed a Gel'fand-Tsetlin basis $\{e_j^i \otimes f_t\}$, and since the simple modules over $A_0 = k$ are one-dimensional, a basis of the A_n-module M has been constructed. It is convenient to number the elements of a basis by paths in a certain graded multigraph. The vertices of its ith level are the simple modules over A_i, and there are edges connecting vertices of the ith and $(i+1)$th levels if there is an inclusion of the corresponding modules; the multiplicity of an edge is s_i (see above). In this case, the set of elements of a Gel'fand-Tsetlin basis is the set of all paths in this graph leading from the initial to the final vertex. The graph of the simple modules of a locally semisimple algebra is called its Bratteli diagram.

EXAMPLE. We will construct the Bratteli diagram of $\text{Alg}\{\sigma_1, \sigma_2, \ldots\}$, where

$$\sigma_i^2 = e, \quad \sigma_i \sigma_{i+1} + \sigma_{i+1} \sigma_i = 0, \quad \sigma_i \sigma_{i+k} = \sigma_{i+k} \sigma_i, \quad k \geq 2, \quad i \geq 1.$$

We begin with the simple modules over $A_1 = \text{Alg}\{\sigma_1\} = k^2$. There are two and they are one-dimensional: $\pi_1: \sigma_i \to 1$, $\pi_2: \sigma_1 \to -1$,

There is only one simple module over $A_2 = \text{Alg}\{\sigma_1, \sigma_2\}$: it is easy to see that as an A_1-module it is a direct sum of π_1 and π_2 with multiplicity 1. We obtain the graph

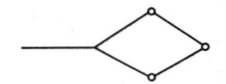

Since σ_3 commutes with σ_1, it follows that in simple modules over A_3 the eigenspaces for σ_3 are the same as for σ_1, hence there are two possibilities, i.e. we obtain the graph

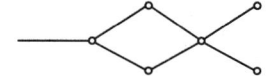

In view of the stationary property, this picture can be continued, and we obtain the graph

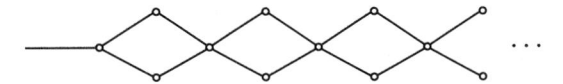

Obviously, $A_{2^n} = M_{2^n}\mathscr{k}$ and $A_{2n+1} = M_{2^n}\mathscr{k} \oplus M_{2^n}\mathscr{k}$. This algebra has about ten interpretations (a spinor algebra, Grassmann algebra, tensor product of second-order matrices, etc.).

To construct the graph of simple modules from relations defining a local algebra means to give its complete harmonic analysis; this, as a rule, is a complicated problem. The inverse problem is to synthesize modules from a graded graph. This construction is carried out below for the symmetric group.

3. Local representations and graded graphs

Suppose a local stationary algebra $A_n = \text{Alg}\{\sigma_1, \ldots, \sigma_n\}$ is given. We will show how to construct modules over this algebra.

Consider an $\mathbb{N}$-graded multigraph Γ. The set of nth level vertices can be either finite or infinite (even continual), but we will assume finiteness of intervals, i.e. finiteness of the number of edges (and vertices) between any vertices. We also assume that edges connect vertices of adjacent levels. Consider the vector $\mathscr{k}$-space $F_n(\Gamma)$ over the set of all paths in Γ leading from the 0th level vertices to the nth level vertices, and fix this basis of paths. Suppose $F_n(\Gamma)$ is an A_n-module; we will call it local if the action of the generator σ_i sends each path (more precisely, the basis element corresponding to it) into a linear combination of paths that may differ from it only in the edges lying between the $(i-1)$th and $(i+1)$th levels:

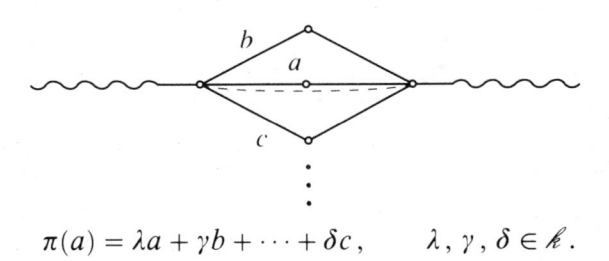

$$\pi(a) = \lambda a + \gamma b + \cdots + \delta c, \qquad \lambda, \gamma, \delta \in \mathscr{k}.$$

LEMMA. *The construction of a Gel'fand-Tsetlin basis (see above) gives a realization of a module in the form of a local module.*

PROOF. We have seen above that a Gel'fand-Tsetlin basis can be enumerated by the paths in a graded graph. Consider the action of a generator σ_i. Since it commutes with $\sigma_1, \ldots, \sigma_{i-2}$, the generator σ_i reduces each primary A_{i-2}-module, hence σ_i does not alter the beginning of a path up to the $(i-1)$th level vertices. By the same reasoning, σ_i preserves the end of a path from the $(i+1)$th level vertices on.

We see that the fact that a module is local is not a restriction on the module. Graphs defining a local A-module will be called A-admissible.

The problem of constructing modules over a local (but not necessarily stationary) algebra is now reduced to describing graded graphs and the actions of (one!) generator on all intervals of length 2. Actually it suffices to describe admissible intervals of length 2, but to do this it is necessary to analyze intervals of length 3, since it is necessary to verify the commutational relation between σ_i and σ_{i+1} on them. Thus the problem of constructing $\mathrm{Alg}\{\sigma_1, \ldots, \sigma_{i+1}\}$-modules from a graph reduces to that of describing $\mathrm{Alg}\{\sigma_i, \sigma_{i+1}\}$-modules and synthesizing global modules from them. Both problems are nontrivial.

As an example, we will construct the modules of the symmetric group (i.e. the Weyl group of the series A_n). Recall that the relations here are as follows:

$$\sigma_i^2 = 1, \quad i = 1, \ldots,$$
$$(\sigma_i \sigma_{i+1})^3 = 1, \quad i = 1, \ldots,$$
$$(\sigma_i \sigma_{i+k})^2 = 1, \quad i = 1, \ldots, \quad k \geq 2.$$

Since the generators have order 2, we need only consider graphs in which each interval of length 2 has at most two paths:

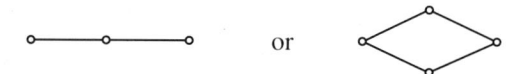

o—o—o   or   ◇

Among the various intervals of length 3 that can be assembled from them we first consider two: 1) a Boolean algebra with three elements; 2) a plane lattice with two cells (see the figure).

Suppose some action of a generator has already been defined on the intervals of length 2 between the first and third levels. How can the action of the next generator be defined on the intervals between the second and fourth levels so that the commutational relation is satisfied?

EXAMPLE. Consider an interval (of the plane lattice). The corresponding space is $\mathbb{R}^3$ (there are three paths between I and IV). If the action of σ_1 is given, then, by what has preceded, it has the following matrix (in the basis of paths):

$$\begin{pmatrix} \cos\alpha & \sin\alpha & 0 \\ \sin\alpha & -\cos\alpha & 0 \\ 0 & 0 & \pm 1 \end{pmatrix};$$

0   I II   III   IV

LEMMA 1. *The action of σ_2 turning $\mathbb{R}^3$ into a local module over $\mathbb{R}(S_3)$ is*

$$\begin{pmatrix} \pm 1 & 0 & 0 \\ 0 & \cos\alpha' & \sin\alpha' \\ 0 & \sin\alpha' & -\cos\alpha' \end{pmatrix},$$

where $\sec\alpha' = \sec\alpha + 1$, i.e. $\cos\alpha' = \cos\alpha/(1+\cos\alpha)$.

(The verification is direct.)

The case of a Boolean algebra is more complicated. This example is fundamental:

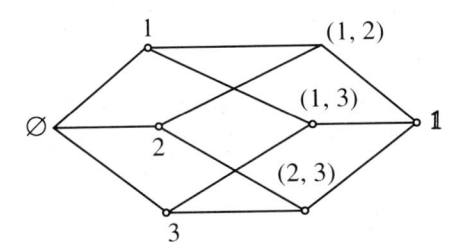

We first form the set of paths in the three-flank Boolean algebra that connect $\varnothing$ and $\mathbf{1}$; there are six such paths: $(\varnothing, 1, (1,2), (1,2,3) = \mathbf{1})$, etc. We define in $\mathbb{R}^6$ a basis enumerated by these paths (flags):

$$e_1 = (1,(1,2)), \quad e_2 = (2,(1,2)), \quad e_3 = (2,(2,3)),$$
$$e_4 = (3,(2,3)), \quad e_5 = (3,(1,3)), \quad e_6 = (1,(1,3))$$

(we have not indicated the common initial point $\varnothing$ or the common terminal point $\mathbf{1}$).

LEMMA 2. *Suppose in $\mathbb{R}^6$, with the above basis, the generator σ_1 ($\sigma_1^2 = e$) acts as a reflection in each of the two-dimensional subspaces $L_1 = \mathrm{Lin}(e_1, e_2)$, $L_2 = \mathrm{Lin}(e_3, e_4)$, $L_3 = \mathrm{Lin}(e_5, e_6)$ and the matrix in the basis (e_{2i-1}, e_{2i}) of L_i, $i = 1, 2, 3$, has the form $\begin{pmatrix} \rho_i & \sqrt{1-\rho_i^2} \\ \sqrt{1-\rho_i^2} & -\rho_i \end{pmatrix}$, and σ_2 acts in $M_1 = \mathrm{Lin}(e_1, e_6)$, $M_2 = \mathrm{Lin}(e_3, e_2)$, $M_3 = \mathrm{Lin}(e_5, e_6)$ by the matrices $\begin{pmatrix} \nu_i & \sqrt{1-\nu_i^2} \\ \sqrt{1-\nu_i^2} & -\nu_i \end{pmatrix}$, $i = 1, 2, 3$, with respect to the basis in parenthesis. Then σ_1, σ_2 define a representation of S_3 (i.e. $(\sigma_1\sigma_2)^3 = 3$) if and only if*

a) $\nu_1 = \rho_2$, $\nu_2 = \rho_3$, $\nu_3 = \rho_1$;
b) $\rho_1\rho_2 + \rho_2\rho_3 + \rho_3\rho_1 = 0$ (or $\rho_1^{-1} + \rho_2^{-1} + \rho_3^{-1} = 0$ when $\rho_i \neq 0$).

This can be proved by a direct (but rather tiresome) calculation.

REMARK. It is convenient to interpret a Boolean algebra as the 1-skeleton of a three-dimensional cube, with σ_1 acting in the two-dimensional faces containing the vertex $\varnothing$ and σ_2 in the two-dimensional faces containing the vertex $\mathbf{1}$, where condition a) means that the actions agree on opposite faces. Condition b) has a more complicated meaning.

Thus the first step has been completed: we have described all $\mathrm{Alg}(\sigma_1, \sigma_2) = S_3$-modules (for two types of intervals of length 3, which basically account for all admissible intervals of length 3; see below). We can now begin the synthesis.

THEOREM. *For a graph to be admissible it is necessary and sufficient that* 1) *any 3-interval be admissible, and that* 2) *different ways of including any given 2-interval in 3-intervals be compatible in the natural sense.*

Indeed, if the action of the generator σ_1 is defined on the "first level," then by induction we define the action of σ_2, etc.

We add without elaboration that the admissible intervals of length 3 (for the group S_3) are exhausted, apart from the above-mentioned two examples of a Boolean algebra and a plane lattice, by three trivial ones:

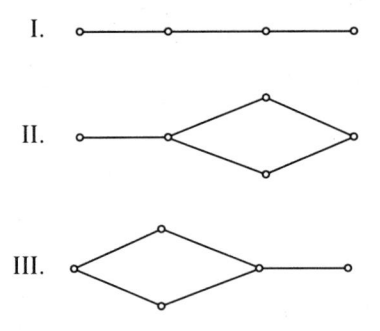

I.

II.

III.

The actions of σ_1 and σ_2 in the space of paths ($\mathbb{R}^1$, $\mathbb{R}^2$, $\mathbb{R}^2$ respectively) are obvious:

$$\sigma_1^{\mathrm{I}} = \sigma_2^{\mathrm{I}} = \pm 1, \qquad \sigma_1^{\mathrm{II}} = \pm 1, \qquad \sigma_2^{\mathrm{II}} = \begin{pmatrix} 0 & 1 \\ 1 & 0 \end{pmatrix};$$

$$\sigma_1^{\mathrm{III}} = \begin{pmatrix} 0 & 1 \\ 1 & 0 \end{pmatrix}, \qquad \sigma_2^{\mathrm{III}} = \pm 1.$$

The choice of $\pm$ is determined by a further stipulation.

These assertions provide an outline of a new construction of the whole representation theory of the symmetric group. Up to now (1987) this theory has been expounded, even in the most modern and the best monographs, in essentially the same form as it was by Young at the beginning of this century (although methodical refinements have completely altered its external appearance and made it more accessible). However, this method of exposition does not answer the basic question (see the epilogue to [8]), "Why Young diagrams?" The method suggested above provides an answer (albeit a partial one) to this question and explains the appearance of Young diagrams not *a posteriori*, as is customary, but *a priori*: the Young lattice is a universal lattice, where only intervals of the indicated five types are encountered.

THEOREM. *The infinite Young lattice (i.e. the lattice of finite ideals of $\mathbb{Z}_+ \oplus \mathbb{Z}_+$) defines a unique S_∞-module in view of the construction described above.*

PROOF. First of all, each interval of length 3 in the Young lattice is S_3-admissible, as noted above. But then it suffices to observe that the first generator acts in $\mathbb{R}^2$, realized as the space of paths in the first level of the graph

$$\begin{pmatrix} +1 & 0 \\ 0 & -1 \end{pmatrix}$$

Then Lemmas 1 and 2 enter the game, these enabling us to extend the actions of the generators in a unique way from one level to the next:

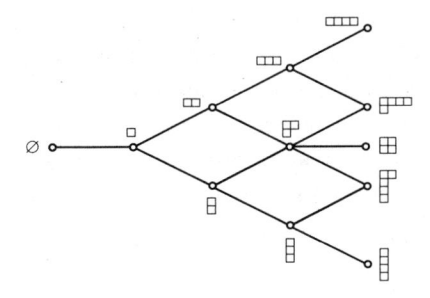

We will now give examples of other graphs admissible for the symmetric group.

1) The Pascal triangle

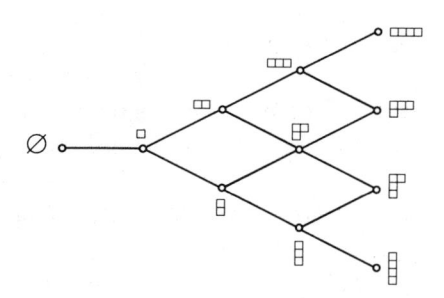

Each interval of length 3 is admissible. This is a subgraph of the Young graph, namely, the graph of diagrams with one arm and leg:

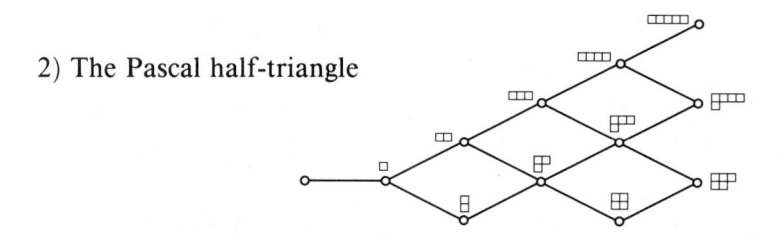

2) The Pascal half-triangle

This is the graph of two-row Young diagrams.
3)

This multigraph describes the natural action of S_∞ in the space $\mathbb{R}^2 \otimes \mathbb{R}^2 \otimes \cdots$.

4) Consider the Boolean algebra of all subsets of the finite set $(1, \ldots, n)$ of n elements. A path between $\varnothing$ and $\mathbf{1}$ is a flag or permutation. Therefore the vector space over the set of paths has dimension $n!$. In this space there acts, in accordance with our recipe, a representation of S_n (isomorphic to the regular representation), whose parameters are the $n-1$ numbers p_k, $0 \leq p_k \leq 1$, $k = 1, \ldots, n$, and the action in the two-dimensional subspace over the paths $(i, j, a, b, \ldots, c)$ and $(j, i, a, b, \ldots, c)$ is given by

$$\begin{pmatrix} \rho_{ij} & \sqrt{1 - \rho_{ij}^2} \\ \sqrt{1 - \rho_{ij}^2} & -\rho_{ij} \end{pmatrix},$$

where $\rho_{ij} = p_{1j} - p_{1i}$, $p_{1k} = p_k$.

The action of $\sigma_2, \ldots, \sigma_{n-1}$ is defined by the same formulas, but in shifted two-dimensional subspaces. The previous argument shows that these formulas define a representation of S_n. An analogous construction can be carried out for S_∞.

This method can be used for other stationary local algebras. We hope to use it for Lie algebras ($GL(\infty)$, $SU(\infty)$, etc.). It is reminiscent of the construction of Markov chains from an initial state and a transition operator. We constructed above a "deterministic" Markov chain, in the non-Abelian case, and it is nontrivial. (These are not the non-Abelian Markov chains that L. Accardi[2] and others have in mind, at least according to the definition, but they have connections with the latter.)

4. The algebra of a braid group. Knot theory

In 1927 E. Artin defined a braid group:

$$\sigma_j \sigma_i = \sigma_j \sigma_i, \qquad |i - j| \geq 2,$$
$$\sigma_i \sigma_{i+1} \sigma_i = \sigma_{i+1} \sigma_i \sigma_{i+1}, \qquad i = 1, \ldots, n, \quad n = 1, \ldots.$$

Every closed braid (the strings of the same level are joined) form a link and, in particular, a knot, if there is one component.

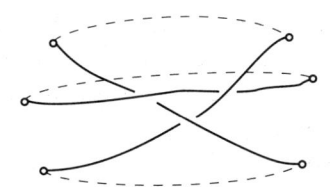

In 1935 A. A. Markov proved that the homotopic classification of links reduces to explicitly giving the partition of a braid group with respect to the equivalence defined by 1) conjugation in the group, and 2) the relation $g\sigma_n = g$, where $g \in B_{n+1}$. Thus the classification problem has been made completely algebraic. However, because of the lack of a well-developed representation theory of braid groups this theorem has not been successfully used until recently. There was a breakthrough in 1983, when Jones [3], [4] observed that the Hecke algebra is a quotient algebra of $\mathbb{R}(B_n)$ and that the Hecke algebra has a one-parameter family of characters satisfying the Markov condition. Thus Jones constructed link invariants, which turned out to be new, and this revived the entire problem, giving rise to a torrent of papers on the subject; there is now a whole series of invariants of this type. In this section we will reproduce a result of our work with Kerov [9], where Jones's result is obtained from an earlier one of Thoma [10] and the authors [11] on the complete list of characters of the group S_∞ and traces on the Hecke algebra. The Jones characters can easily be picked out from this list.

DEFINITION. A trace of a local stationary algebra is called Markov[3] if it satisfies the condition

$$\chi(h\sigma_n) = w\chi(h),$$
$$h \in \mathrm{Alg}\{\sigma_1, \ldots, \sigma_{n-1}\}, \qquad w \in \mathbb{C}.$$

---

[2] See, for example, L. Accardi, Funktsional. Anal. i Prilozhen. **9** (1975), 1-8; MR **52** #4845.—Translator.

[3] In honor of A. A. Markov, Jr. (see above). This trace is Markov in the sense of the theory of Markov processes (see [1]), named for A. A. Markov, Sr.

A character of a braid group is Markov if
$$\chi(h\sigma_n^{\pm 1}) = w_\pm \chi(h), \qquad h \in B_{n-1}, \qquad w \in \mathbb{C}.$$

From a Markov character of a braid group we can construct the link invariant
$$I_\chi(L_b) = m^s \chi(b)/w^{n-1},$$
where $m = (w_-/w_+)^{1/2}$, $w = (w_- w_+)^{1/2}$, s is the sum of the exponents of the generators occurring in the braid b, L_b is the link obtained by closing b, and $w_\pm$ are parameters.

The method for obtaining Markov characters utilizes the fact that under the homomorphism $\mathbb{C}(B_n) \to H_n(q)$ a Markov trace on the Hecke algebra $H_n(q)$ is a Markov character on B_n. Therefore the problem has been reduced to describing the Markov traces on $H_n(q)$. Since $H_n(q)$ and $\mathbb{C}(S_n)$ are isomorphic for all q different from 0 and roots of unity, it suffices to pass to generators of the Hecke algebra in the formulas of [10] and [11] for the characters of the symmetric group.

We will give these formulas. The parameters of the characters are two sequences $\alpha = (\alpha_1, \ldots)$, $\beta = (\beta_1, \ldots)$, where
$$\alpha_1 \geq \alpha_2 \geq \cdots \geq 0, \quad \beta_1 \geq \beta_2 \geq \cdots \geq 0, \quad \sum_{i=1}^\infty (\alpha_i + \beta_i) \leq 1.$$

If σ_i are the generators of the Hecke algebra $H_\infty(q)$, then a character is completely determined by its values on the elements of the form $c^{(n)} = \sigma_1 \sigma_2 \cdots \sigma_n$, $n = 1, \ldots$. If $\chi^{\alpha, \beta}$ denotes a character and $t_n(\alpha, \beta, q) = \chi^{\alpha, \beta}(c^{(n)})$ is its value on $c^{(n)}$, then the generating function $T_{\alpha, \beta}(z) = 1 + (q-1) \sum_{n=1}^\infty t_n z^n$ has the form
$$T_{\alpha,\beta}^{(q)}(z) = e^{\gamma(q-1)z} \prod_i \frac{(1-\alpha_i q z)}{1-\alpha_i z} \cdot \frac{1+\beta_i z}{1+q\beta_i z},$$
$$\gamma = 1 - \sum_i (\alpha_i + \beta_i) \geq 0.$$

When $q = 1$, the generators σ define the symmetric group, and the values are the following:
$$\chi^{\alpha,\beta}(c^{(n)}) = t_n(\alpha, \beta, 1) = \sum_i \alpha_i^n + (-1)^{n+1} \sum_i \beta_i^n,$$
where $c^{(n)}$ is an n-cycle. These formulas are studied in detail in [11].

It remains to prove the following fact.

THEOREM. *A character of the algebra $H_\infty(q)$ is Markov if and only if the α- and β-sequences are geometric progressions with ratio $1/q$ and are infinite for $\alpha_1, \beta_1 \neq 0$, and finite otherwise.*

From this assertion it is easy to obtain the Jones invariants.

Bibliography

1. A. M. Vershik and S. V. Kerov, *Locally semisimple algebras*, in: Contemporary Problems of Mathematics, Vol. 26, VINITI, Moscow, 1985, pp. 3–56. (Russian)
2. A. Connes, Sém. N. Bourbaki, 1985, Exp. 647.
3. V. F. R. Jones, *Index for subfactors*, Invent. Math. **72** (1983), 1–25.
4. ___, *Hecke algebra representations of braid groups and link polynomials*, Ann. Math. **126** (1987), 333–388.
5. N. Bourbaki, *Groupes et algèbres de Lie*, Ch. IV–VI, Hermann, Paris, 1968.
6. D. Kazhdan and G. Lusztig, *Representations of Coxeter groups and Hecke algebras*, Invent. Math. **53** (1979), 165–184.
7. G. James and A. Kerber, *The representation theory of the symmetric group*, Addison-Wesley, New York, 1981.
8. G. James, *The representation theory of the symmetric groups*, Lecture Notes in Math., Vol. 682, Springer-Verlag, Berlin and New York, 1978.
9. A. M. Vershik and S. V. Kerov, *Characters and realizations of representations of an infinite-dimensional Hecke algebra, and knot invariants*, Dokl. Akad. Nauk SSSR **301** (1988), 777–780; English transl. in Soviet Math. Dokl. **38** (1989), 134–137.
10. E. Thoma, *Die unzerlegbaren, positiv-definiten Klassenfunktionen der abzählbar unendlichen, symmetrischen Gruppe*, Math. Z. **85** (1964), 40–61.
11. A. M. Vershik and S. V. Kerov, *Asymptotic theory of characters of the symmetric group*, Funktsional. Anal. i Prilozhen. **15** (1981), no. 4; English transl. in Functional Anal. Appl. **15** (1981).
12. A. M. Vershik, *Local algebras and a new version of Young's orthogonal form*, in: Topics in Algebra, Banach Center Publications, vol. 26, part 2, pp. 467–473, PWN, Warsaw, 1990.

Translated by G. A. KANDALL

The Volume of Polyhedra on a Sphere and in Lobachevsky Space

UDC 514.132

È. B. VINBERG

The calculation of volumes of non-Euclidean (three-dimensional) polyhedra is an entirely novel subject, having nothing in common with calculation of volumes of Euclidean polyhedra or areas of non-Euclidean polygons. Strange as it may seem, almost everything that is known at the present time about the subject is contained in the half-forgotten papers of Lobachevsky [1], [2] and Schläfli [3]. In the first two papers the topic is polyhedra in Lobachevsky space, in the third spherical polyhedra. Coxeter [4] compared these papers from the point of view of analytic continuation of volume formulas. After "reduction to a common denominator" it became clear that the papers of Lobachevsky and Schläfli do not overlap but on the contrary complement each other well. Nevertheless Coxeter's paper did not attract proper notice to the subject perhaps because there was no particular need. However, in recent years, thanks to the discovery by Thurston of the connection between Lobachevskian geometry and the topology of three-dimensional manifolds, interest in the calculation of volumes of non-Euclidean polyhedra has grown sharply. Namely for this reason, in the article of Milnor [5], dedicated to the 150th anniversary of Lobachevskian geometry, there was a supplement on calculation of volumes in which the investigations of Lobachevsky were mentioned and an independent derivation was given of one simple consequence of a formula of Lobachevsky.

In the present lecture I give a modernized account of the basic results of Lobachevsky and Schläfli, incorporating improvements proposed by Coxeter and Milnor and adding a formula of my own for the volume of a pyramid with vertex at infinity.

The material on non-Euclidean geometry having no direct connection with

---

1980 *Mathematics Subject Classification* (1985 *Revision*). Primary 51M10, 51M20, 51M25; Secondary 52A55.

calculation of volumes may be found, for example, in [6], [7], [8].

The metrics for the sphere and Lobachevsky space are assumed to be normalized so that the curvature χ equals ± 1. All formulas containing the symbol χ reduce, for $\chi = 0$, to correct formulas of Euclidean geometry.

1. As is known, the area of a non-Euclidean m-gon P with angles $\alpha_1, \ldots, \alpha_m$ is found from the formula

$$\chi \text{ area } P = \sum \alpha_i - \pi(m-2). \tag{1}$$

2. The three-dimensional sphere S^3 and the Lobachevsky space Π^3 can be described uniformly by means of the "vector model". Namely the sphere S^3 is the hypersurface in the Euclidean vector space E^4 defined by the equation $(x, x) = x_0^2 + \cdots + x_3^2 = 1$. A Riemannian metric on the sphere is induced by the Riemannian metric $ds^2 = dx_0^2 + \cdots + dx_3^2$ of the space E^4.

Likewise, the space Π^3 is the hypersurface in the pseudo-Euclidean space $E^{3,1}$ defined by the equation $(x, x) = -x_0^2 + x_1^2 + x_2^2 + x_3^2$, $x_0 > 0$. A Riemannian metric on the space Π^3 is induced by the pseudo-Riemannian metric $ds^2 = -dx_0^2 + dx_1^2 + dx_2^2 + dx_3^2$ on the space $E^{3,1}$.

Motions of the space S^3 (Π^3) in this model are induced by orthogonal (pseudo-orthogonal) transformations of the space E^4 ($E^{3,1}$), and lines and planes are represented by intersections of S^3 (Π^3) with subspaces of the space E^4 ($E^{3,1}$). In particular, every plane can be represented in the form $H_e = \{x : (x, e) = 0\}$ where e is a vector satisfying the condition $(e, e) = 1$. The angle between hyperplanes H_e and H_f (if they intersect) is found from the formula

$$\cos \widehat{H_e H_f} = -(e, f). \tag{2}$$

The distance $\rho(x, y)$ between points x and y in the vector model is found in the case of the sphere from the formula

$$\cos \rho(x, y) = (x, y), \tag{3}$$

and in the case of Lobachevsky space from the analogous formula

$$\cosh \rho(x, y) = -(x, y). \tag{4}$$

3. Since a basis of the space E^4 ($E^{3,1}$) is uniquely determined, up to an orthogonal (pseudo-orthogonal) transformation, by its Gram matrix, it follows from formula (2) that a tetrahedron in the space S^3 (Π^3) is uniquely determined, up to a motion, by its dihedral angles. Note that a tetrahedron in the Euclidean space E^3 is determined by its dihedral angles only up to a similarity transformation.

For each vector $\alpha \in \mathbb{R}^6$ we denote by $T(\alpha)$ the tetrahedron of the space S^3, E^3, or Π^3, if there is such, whose dihedral angles, ordered in some manner, coincide with the coordinates of the vector α. With the help of the vector model it is not difficult to show that there exists a two-valued analytic function F defined in some domain of the space $\mathbb{C}^6$ containing the 6-tuples of coordinates of the dihedral angles of all tetrahedra of the spaces S^3, E^3, $Л^3$, and satisfying the conditions

$$F(\alpha) = \begin{cases} \pm \operatorname{vol} T(\alpha), & \text{if } T(\alpha \subset S^3, \\ 0 & \text{if } T(\alpha) \subset E^3, \\ \pm i \operatorname{vol} T(\alpha), & \text{if } T(\alpha) \subset Л^3. \end{cases} \quad (5)$$

This permits one, by means of analytic continuation, to obtain formulas for the volumes of polyhedra in the space $Л^3$ from formulas for the volumes of spherical polyhedra, and vice versa.

4. Formulas for volumes of certain bodies can be obtained by integration of volume elements in an appropriate coordinate system. For these purposes special interest is presented by a "cylindrical" coordinate system (Figure 1), in which a point x is given by its distance r from the axis l, the angle φ between the perpendicular dropped from x to l and some fixed plane passing through l, and the coordinate t of the foot of this perpendicular. The coordinate lines are rays perpendicular to the axis l lying in planes perpendicular to it, and lines equidistant from the axis l in planes passing through the axis. These lines are mutually perpendicular at each point.

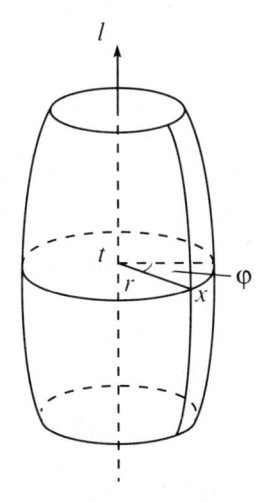

FIGURE 1

Since the circumference of a circle of radius r on the sphere (or in the Lobachevsky plane) equals $2\pi \sin r (2\pi \sinh r)$, the length of a circular arc of radius r with central angle $d\varphi$ equals $\sin r\, d\varphi (\sinh r\, d\varphi)$. Likewise, the length of an equidistant arc with projection dt on the baseline equals $\cos r\, dt (\cosh r\, dt)$. Hence the volume element in our cylindrical coordinate system has the form

$$dV = \begin{cases} \sin r \cos r\, dr\, d\varphi\, dt & \text{in } S^3, \\ \sinh r \cosh r\, dr\, d\varphi\, dt & \text{in } Л^3. \end{cases} \quad (6)$$

5. An elliptic wedge here denotes an infinitesimally narrow triangular prism whose lateral edge h is perpendicular to both bases (Figure 2). By integration in a cylindrical coordinate system whose axis runs along the edge h, one can obtain the following formula for the volume of an elliptic wedge:

$$\chi dV = \tfrac{1}{2}(h - g\cos\theta)\,d\varphi, \tag{7}$$

where $d\varphi$ is the dihedral angle at the edge h, g is the other lateral edge, and θ is the angle between the edges h and g.

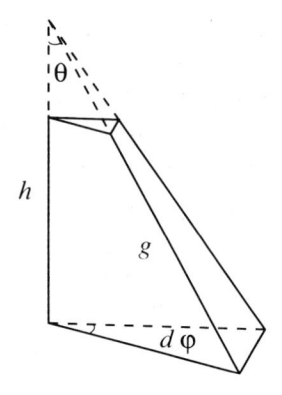

FIGURE 2

By integration of equality (7) with respect to φ, one obtains the following formula for the volume of a right circular cone: $\chi V = \pi(h - g\cos\theta)$, where h is the altitude, g the generator, and θ the angle between them.

6. A hyperbolic wedge here denotes an infinitesimally thin triangular prism whose lateral edge dh is perpendicular to the bases (Figure 3). By integration in the cylindrical coordinate system whose axis runs along the edge dh, one obtains the following formula for the volume of a hyperbolic wedge:

$$dV = \tfrac{1}{2}a\,dS, \tag{8}$$

where a is the side of the base opposite dh and dS is the area of the wedge cross-section by the plane passing through dh and perpendicular to a.

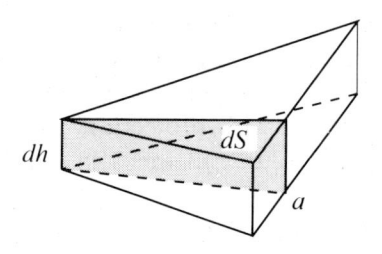

FIGURE 3

It is interesting that formula (8) has the same form on both the sphere and the Lobachevsky plane. It also holds in the Euclidean plane.

7. Starting from formula (8), we get Schläfli's formula for the volume differential of a non-Euclidean polyhedron

$$\chi d\,\text{vol}\,P = \tfrac{1}{2}\sum a_i\,d\alpha_i. \tag{9}$$

Here it is assumed that the polyhedron P is deformed in such a way that it retains its combinatorial structure, the summation runs over all its edges a_i, and α_i denotes the dihedral angle at the edge a_i.

It is not difficult to see that both sides of equation (9) are additive in the sense that if the polyhedron P decomposes into several polyhedra and this decomposition is deformed together with P, then each expression is the sum of the corresponding expressions for the polyhedra of the decomposition. Thus it suffices to demonstrate formula (9) for a tetrahedron.

Furthermore, it suffices to consider a deformation of the tetrahedron in which only one dihedral angle changes. Such a deformation consists of a displacement of one plane H bounding the tetrahedron T along a straight line m that contains an edge not lying in H (Figure 4). In this process, only the angle α between the plane H and the plane of the other face not containing m is changed. The increment in the volume of the tetrahedron equals the volume of the triangular prism included between the plane H and its new position H'; this prism could be regarded, in first approximation, as a hyperbolic wedge with edge lying on a line l perpendicular to H. The cross-section of this wedge by a plane running through l and perpendicular to the opposite side a of the wedge base (coinciding with an edge of the tetrahedron T) is a quadrangle with two right angles, the two remaining angles being equal to $\pi - \alpha$ and $\alpha' = \alpha + d\alpha$. By formula (1) the area of this quadrangle multiplied by χ is equal to $d\alpha$, and by formula (8) we get $\chi d(\text{vol}\,T) = (1/2)a\,d\alpha$, q.e.d.

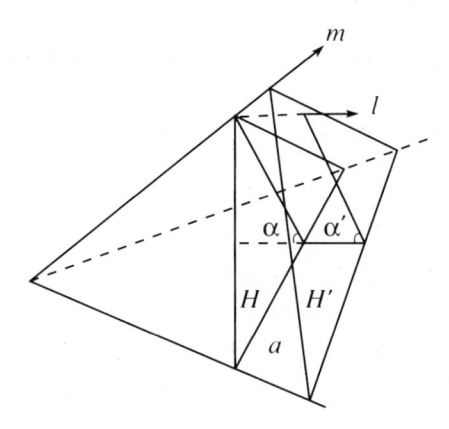

FIGURE 4

8. A tetrahedron $ABCD$ is called birectangular if edge AB is perpendicular to the plane BCD and edge CD to the plane DAB (Figure 5). When calculating volumes it is possible, in principle, to restrict oneself to birectangular tetrahedra because every tetrahedron can be decomposed into six birectangular ones (Figure 6) by dropping perpendiculars from one vertex to the plane of the opposite face and to the lines bounding that face. (If some of the perpendiculars lie outside the tetrahedron, we do not obtain a decomposition in the usual sense but rather an "algebraic sum" in which some terms have a negative sign.)

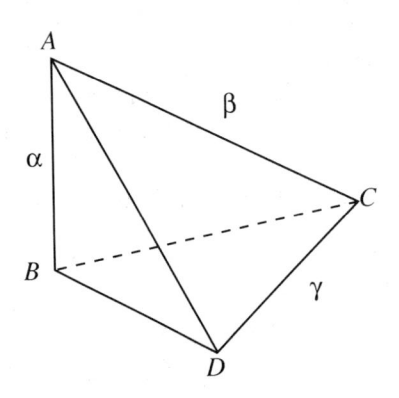

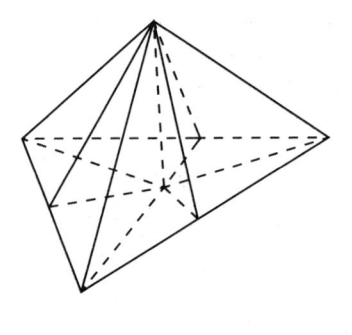

FIGURE 5

FIGURE 6

Of the six dihedral angles of a birectangular tetrahedron three are right angles. The other three are denoted by α, β, and γ as in Figure 5. A birectangular tetrahedron is completely determined by these three angles. In particular, the edges a, b, c corresponding to them are found in the case of the sphere from the formulas

$$\tan a \tan d = \tan b \tan\left(\frac{\pi}{2} - \beta\right) = \tan c \tan \gamma = \frac{\sqrt{\Delta}}{\cos\alpha \cos\gamma}, \qquad (10)$$

and in the case of Lobachevsky space from the similar formula

$$\tanh a \tan d = \tanh b \tan\left(\frac{\pi}{2} - \beta\right) = \tanh c \tan \gamma = \frac{\sqrt{-\Delta}}{\cos\alpha \cos\gamma}, \qquad (11)$$

where in both cases

$$\Delta = \sin^2\alpha \sin^2\gamma - \cos^2\beta \qquad (12)$$

is the Gram determinant of the system of unit vectors orthogonal to the faces of the tetrahedron in the vector model.

We dwell in greater detail on the case of Lobachevsky space. Following an idea of Lobachevsky, in this case we define an angle δ by the formula

$$\tan \delta = \frac{\sqrt{-\Delta}}{\cos\alpha \cos\gamma}. \qquad (13)$$

Elementary computations show that
$$a = \frac{1}{2} \ln \frac{\sin(\alpha + \delta)}{\sin(\alpha - \delta)}, \qquad b = \frac{1}{2} \ln \frac{\sin(\pi/2 - \beta + \delta)}{\sin(\pi/2 - \beta - \delta)},$$
$$c = \frac{1}{2} \ln \frac{\sin(\gamma + \delta)}{\sin(\gamma - \delta)}. \qquad (14)$$

9. For the calculation of the volume of a birectangular tetrahedron $T = ABCD$ in Lobachevsky space, we will contract it to an interval or a point in such a way that in the contraction process the tetrahedron remains birectangular and the angle δ is unchanged. This can be accomplished, for example, by a rotation of the plane ADC about the line l passing through the vertex A perpendicular to the face ABD (Figure 7): by such a deformation neither the edge a nor the angle α is changed and hence neither is the angle δ.

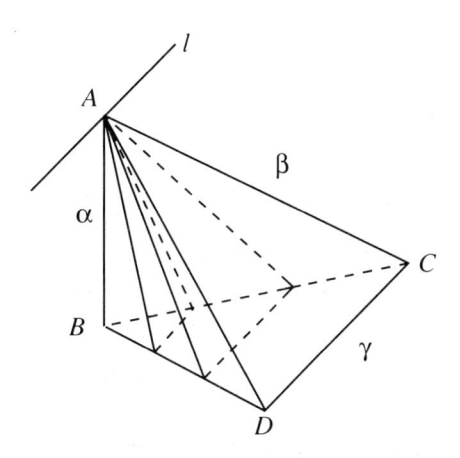

FIGURE 7

Schläfli's formula (9) in the case under consideration takes the form
$$d \operatorname{vol} T = -\tfrac{1}{2}(a\,d\alpha + b\,d\beta + c\,d\gamma). \qquad (15)$$
If we substitute here the expressions (14) for the edges a, b, c, and if we take into account that $\delta = \operatorname{const}$, we get a differential form with separated variables. Its integral is represented in the form of a linear combination of integrals of the function $\ln \sin \xi$ with various limits.

Following Milnor [5] we shall use the name Lobachevsky function, and the symbol Л, for the function
$$\text{Л}(x) = -\int_0^x \ln |2 \sin \xi|\, d\xi. \qquad (16)$$

This function is associated with the function $L(x) = -\int_0^x \ln \cos \xi\, d\xi$, traditionally called the Lobachevsky function, by the relation $L(x) = \text{Л}(x - (\pi/2)) + x \ln 2$.

The function Л is odd, periodic with period π, equal to zero at the points $n\pi/2$, and achieves a maximum equal to $M \simeq 0.50747$ at the points $n\pi + \pi/6$ and a minimum $-M$ at the points $n\pi - \pi/6$ (Figure 8). It is analytic everywhere except at the points $n\pi$, where its derivative has the value $+\infty$, and admits the identity

$$Л(2x) = 2[Л(x) + Л(x + \pi/2)]. \tag{17}$$

Concerning other properties of the function Л, and in particular its connection with the dilogarithm, see [5], [9].

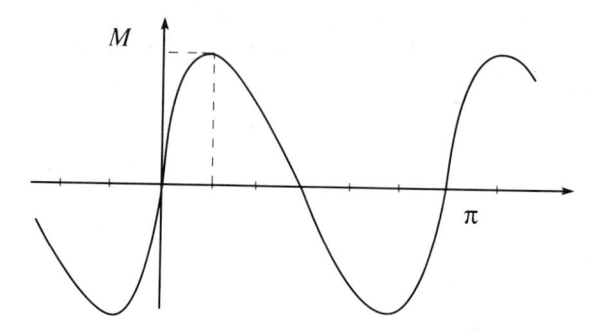

FIGURE 8

By integration of the differential form (15) the following formula of Lobachevsky is obtained for the volume of a birectangular tetrahedron:

$$\text{vol}\, T = \tfrac{1}{4}[Л(\alpha + \delta) - Л(\alpha - \delta) - Л(\tfrac{\pi}{2} - \beta + \delta) \\ + Л(\tfrac{\pi}{2} - \beta - \delta) + Л(\gamma + \delta) - Л(\gamma - \delta) + 2Л(\tfrac{\pi}{2} - \delta)]. \tag{18}$$

It is necessary, it is true, to note that in the original paper of Lobachevsky the angle β, like δ, was expressed by other elements of the tetrahedron; that it is in reality a dihedral angle of the tetrahedron was noted by Coxeter [4]. It would also be interesting to explain the geometric meaning of the angle δ.

By the principle of analytic continuation (5) the formula of Lobachevsky can be carried over to spherical tetrahedra, but in this case the quantity δ will be imaginary (in particular, for $\cos^2 \alpha + \cos^2 \beta + \cos^2 \gamma > 1$, pure imaginary).

10. The formula of Lobachevsky is sufficiently simple, but in order to use it to calculate the volume of an arbitrary polyhedron, it is necessary to decompose the latter into birectangular tetrahedra and calculate their angles, a process, which in general requires the solution of a system of algebraic equations. Therefore, of interest are explicit formulas for the volume of polyhedra in $Л^3$ with infinitely distant vertices.

If p is an infinitely distant vertex of a polyhedron P, then a sufficiently small horosphere with center at the point p intersects P in a Euclidean polygon whose angles equal the corresponding dihedral angles of the polyhedron P. Therefore, if m of the faces converge to the vertex p, then the sum of the angles between them is $\pi(m-2)$.

In particular, if in the birectangular tetrahedron $T = ABCD$ the vertices A and C are infinitely distant, then $\alpha = (\pi/2) - \beta = \gamma = \delta$, and the Lobachevsky formula, in view of (17), takes the form

$$\operatorname{vol} T = \tfrac{1}{2}\Lambda(\alpha). \tag{19}$$

This gives a geometric interpretation for the Lobachevsky function.

We consider now a tetrahedron T (no longer birectangular) all of whose vertices are infinitely distant. The sum of the dihedral angles at any vertex of such a tetrahedron equals π. From this it is not difficult to deduce that the dihedral angles at opposite edges are equal. Consequently, the tetrahedron T is determined by the angles α, β, and γ at a single vertex, for which $\alpha + \beta + \gamma = \pi$. We shall decompose it into six birectangular tetrahedra, as in Figure 6. One of the dihedral angles of each of these six tetrahedra coincides with a dihedral angle of the tetrahedron T. This gives us the opportunity to obtain their volumes by formula (19). As a result one obtains the following formula of Milnor:

$$\operatorname{vol} T = \Lambda(\alpha) + \Lambda(\beta) + \Lambda(\gamma). \tag{20}$$

In Milnor's paper [5] a direct proof of this formula was given without the use of Lobachevsky's formula.

11. We consider a tetrahedron with three infinitely distant vertices. The dihedral angles at the fourth vertex C are denoted by α, β, γ and the others by α_1, β_1, γ_1 as in Figure 9 (infinitely distant vertices are indicated by small circles). From the conditions on the dihedral angles at the infinitely distant vertices it follows that

$$\begin{aligned}\alpha_1 &= \tfrac{1}{2}(\pi + \alpha - \beta - \gamma), \\ \beta_1 &= \tfrac{1}{2}(\pi - \alpha + \beta - \gamma), \\ \gamma_1 &= \tfrac{1}{2}(\pi - \alpha - \beta + \gamma).\end{aligned} \tag{21}$$

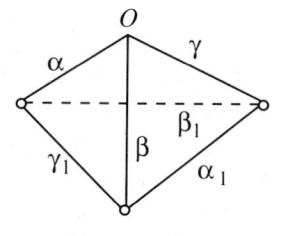

FIGURE 9

For the calculation of the volume of the tetrahedron T we extend its edges containing the vertex O to infinity. We obtain in toto six infinitely distant points which serve as the vertices of an octahedron. The planes of the faces of the tetrahedron T decompose this octahedron into eight tetrahedra, symmetrically located in pairs with respect to their common vertex O (Figure

10). Each of them has three infinitely distant vertices, and their dihedral angles at the vertex O are equal to the dihedral angles of the tetrahedron T or else are adjacent to them. With the aid of formulas (21) all the other dihedral angles of these tetrahedra can be expressed in terms of the dihedral angles of the tetrahedron T.

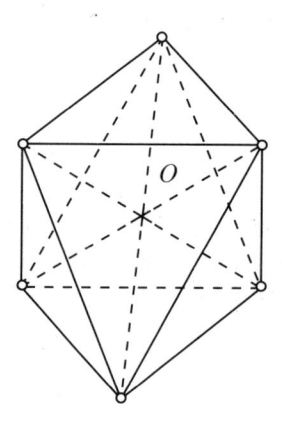

FIGURE 10

Any two adjacent tetrahedra of the decomposition form a tetrahedron all of whose vertices are infinitely distant. The volumes of such tetrahedra can be found by Milnor's formula (20). On the other hand, it is not difficult to find a linear combination of them equal to the volume of the tetrahedron T. For example, if T_1 and T_2 are tetrahedra of the decomposition adjacent to T, and T_2' is the tetrahedron symmetric to T_2, then

$$2\operatorname{vol} T = \operatorname{vol}(T \cup T_1) + \operatorname{vol}(T \cup T_2) - \operatorname{vol}(T_1 \cup T_2').$$

As a result we get the formula

$$\operatorname{vol} T = \tfrac{1}{2}[Л(\alpha) + Л(\beta) + Л(\gamma) + Л(\alpha_1) + Л(\beta_1) \\ + Л(\gamma_1) - Л(\tfrac{1}{2}(\alpha + \beta + \gamma - \pi))]. \quad (22)$$

12. Formulas (20) and (22) are special cases of a formula for the volume of a pyramid with an infinitely distant vertex.

We consider first a quadrangular pyramid P with an infinitely distant vertex C, whose lateral edge CD is an altitude perpendicular to the base, with the sides of the base emanating from D perpendicular to the other two sides. Its dihedral angles at the vertex O are denoted by α, β, and γ as in Figure 11. Then the dihedral angle at the edge CD is equal to $\pi - \gamma$, while all the other dihedral angles are right angles.

To calculate the volume of the pyramid we extend to infinity the sides of the base containing the vertex O. We denote by A and B the infinitely distant points so obtained. The tetrahedron $T = OABC$ is the sum of the pyramid, two birectangular tetrahedra T_1 and T_2 adjacent to it with two infinitely distant vertices each, and a tetrahedron $T_0 = ABCD$ (Figure 12).

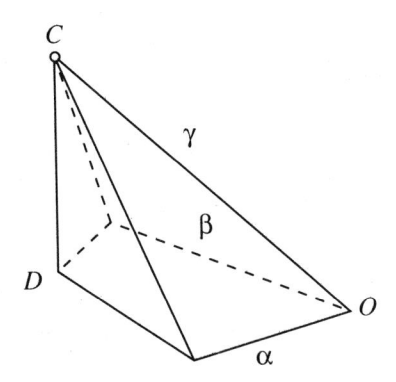

FIGURE 11

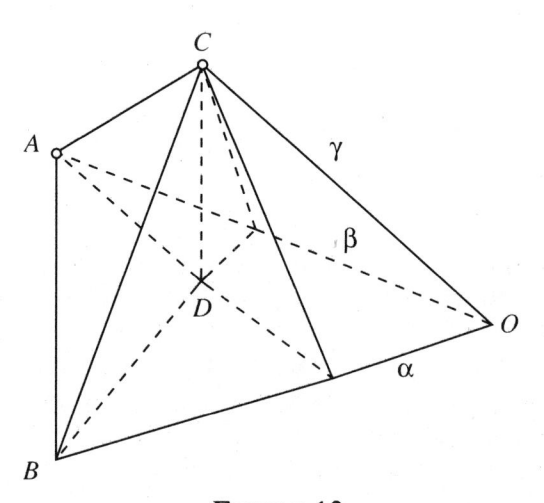

FIGURE 12

The volumes of the tetrahedra T_1 and T_2 can be found by formula (19), and the volumes of the tetrahedra T and T_0 by formula (22). As a result we get the formula

$$\operatorname{vol} P = \tfrac{1}{2} \left[Л(\gamma) + \sum Л(\tfrac{1}{2}(\pi \pm \alpha \pm \beta - \gamma)) \right], \tag{23}$$

where the summation is carried out over all combinations of signs "+" and "−". In the special case $\gamma = \pi/2$ this formula was obtained by Lobachevsky.

An arbitrary n-angular pyramid P with an infinitely distant vertex can be decomposed into n quadrangular pyramids of the above type, by dropping perpendiculars from its vertex to the plane of the base and to the bounding lines of the base (Figure 13). Thus the volume of such a pyramid equals

$$\operatorname{vol} P = \tfrac{1}{2} \sum_{i=1}^{n} \left[Л(\gamma_i) + \sum Л(\tfrac{1}{2}(\pi \pm \alpha_i \pm \alpha_{i+1} - \gamma_i)) \right], \tag{24}$$

where $\alpha_1, \ldots, \alpha_n$ are the dihedral angles of the base, $\alpha_{n+1} = \alpha_1$, and $\gamma_1, \ldots, \gamma_n$ are the dihedral angles of the lateral edges. It would be interesting

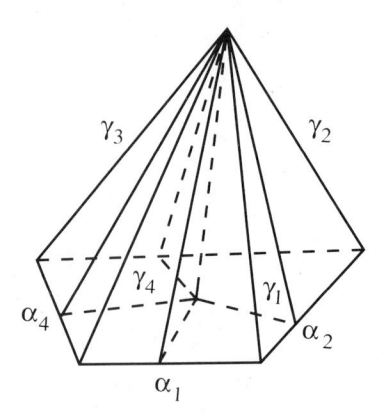

FIGURE 13

to describe all polyhedra with infinitely distant vertices for which there exist formulas of similar type.

13. A decomposition of space into polyhedra is called homogeneous if for any two of its polyhedra P_1 and P_2 there is a motion preserving the decomposition and transforming P_1 into P_2. One of the ways of getting a homogeneous decomposition is related to Coxeter polyhedra—convex polyhedra all of whose dihedral angles are π divided by an integer. If P is a Coxeter polyhedron, then polyhedra obtained from it by successive reflections with respect to its faces form a homogeneous decomposition. Such decompositions are called kaleidoscopic.

We say that one homogeneous decomposition is finer than another if the volume of (any) polyhedron in the first decomposition is less than the volume of a polyhedron in the second. It is known [10], [11] that in any collection of homogeneous decompositions of Lobachevsky space there is a finest decomposition. In particular, there is a finest one among all decompositions. It would be of interest to find this decomposition.

As was proved by the author and O. P. Ruzmanov, the finest of all the kaleidoscopic decompositions is the decomposition determined by the birectangular tetrahedron T_1 with the angles $\pi/4$, $\pi/3$, and $\pi/5$. The volume of this tetrahedron equals $\simeq 0.03589$. The next finest among the kaleidoscopic decompositions apparently is determined by the birectangular tetrahedron T_2 with angles $\pi/3$, $\pi/5$, and $\pi/3$, whose volume equals $\simeq 0.03905$.

The tetrahedron T_2 has a second-order symmetry axis. By using this it is possible to construct a (nonkaleidoscopic) decomposition twice as fine as the previous one. This decomposition is the finest one among all homogeneous decompositions with an arithmetic symmetry group [12]. It is possible that it is the finest among all decompositions.

Among the homogeneous decompositions on polyhedra with infinitely distant vertices, the finest is the kaleidoscopic decomposition determined by the birectangular tetrahedron with angles $\pi/6$, $\pi/3$, and $\pi/3$ [11]. The volume of this tetrahedron equals $\simeq 0.04229$.

Bibliography

1. N. I. Lobachevsky, *Imaginary geometry*, in: Complete Collected Works, Moscow and Leningrad, 1949, vol. 3, pp. 16–70. (Russian)

2. ———, *Application of imaginary geometry to some integrals*, in: Complete Collected Works, Moscow and Leningrad, 1949, vol. 3, pp. 181–294. (Russian)

3. L. Schläfli, *On the multiple integral $\int\int \cdots \int dx\,dy\cdots dz$ whose limits are $p_1 = a_1 x + b_1 y + \cdots + h_1 z > 0$, $p_2 > 0, \ldots, p_n > 0$, and $x^2 + y^2 + \cdots + z^2 < 1$*, Quart. J. Math. **2** (1858), 269–300; **3** (1860), 54–68, 97–108.

4. H. S. M. Coxeter, *The functions of Schläfli and Lobatschevsky*, Quart J. Math. **6** (1935), 13–29.

5. J. Milnor, *Hyperbolic geometry: the first 150 years*, Bull. Amer. Math. Soc. **6** (1982), 9–24.

6. V. F. Kagan, *Foundations of geometry*, Part I, GTTI, Moscow and Leningrad, 1949; Part II, GTTI, Moscow and Leningrad, 1956. (Russian)

7. P. A. Shirokov, *A short outline of the bases of Lobachevskian geometry*, "Nauka", Moscow, 1983. (Russian)

8. M. Berger, *Géométrie*, fasc. 1–5, CEDIC/Fernand Nathan, Paris, 1977, 1978.

9. L. Lewin, *Dilogarithms and associated functions*, London, 1958.

10. W. P. Thurston, *Three-dimensional manifolds, Kleinian groups and hyperbolic geometry*, Bull. Amer. Math. Soc. **6** (1986), 357–382.

11. R. Meyerhoff, *Sphere packings and volumes of hyperbolic 3-spaces*, Comm. Math. Helv. **61** (1986), 271–278.

12. T. Chinburg and E. Friedman, *The smallest arithmetic hyperbolic three-orbifold*, Invent. Math. **86** (1986), 507–528.

Translated by A. VOGT

Integral Geometry on Symmetric Manifolds

UDC 514.765.7

S. G. GINDIKIN

The main subject matter of this lecture is a new derivation of the Plancherel formula for complex semisimple Lie groups based entirely on the ideas of integral geometry. However, this derivation generalizes immediately to other pseudo-Riemannian symmetric manifolds for which the problem of integral geometry is solvable.

Let us recall that the Plancherel formula for complex semisimple Lie groups was obtained in 1950–51 by Gel'fand and Naĭmark [1] and Harish-Chandra [2]. Sufficiently clear proofs of it, based on the regularization of distributions with a parameter, were later obtained in [3]. In 1959 Gel'fand and Graev [4] singled out a problem of integral geometry equivalent to the Plancherel formula: reconstruct the function on a group if its integrals over horospheres—over translations of the maximal unipotent subgroup—are known. The natural plan consisted of including this problem in a more general class of integral-geometry problems by replacing the horospheres by some other submanifolds and devising general procedures for inverting such integral transformations.

In 1967 Gel'fand, Graev, and Shapiro [5] determined the general structure of inversion formulas (the form $\varkappa$) for the case when the integration is carried out over some family of p-dimensional planes in $\mathbb{C}^n$. This made it possible [6] to obtain the Plancherel formula for the group $SL(l, \mathbb{C})$, since its horospheres can be interpreted as planes of dimension $l(l-1)/2$ in $\mathbb{C}^{l^2-1}$. The passage to other groups requires the ability to solve integral-geometry problems with integration over curvilinear submanifolds. Very complete results in this direction were obtained for the case of curves in [7] and [8]. Below, results are presented for problems connected with curvilinear submanifolds of higher dimension that are sufficient for obtaining the Plancherel formula.

1980 *Mathematics Subject Classification* (1985 *Revision*). Primary 53C65, 22E46.

Interest in an integral-geometric derivation of the Plancherel formula is connected with the fact that it must exhibit the geometric structures responsible for the existence of an explicit Plancherel formula, and there is no doubt that the existence of a group action is too high a price for this. We show that this is connected with the existence of a very simple differential-geometric structure on the manifold of horospheres. This structure can naturally be regarded as one of the possible multidimensional variants of structures that arise in the twistor theory of Penrose [9].

Let us emphasize once more the two main problems of our investigations:

(i) obtain a sufficiently general formula in problems of integral geometry, applicable, in particular, to the problem on complex semisimple Lie groups;

(ii) exhibit the geometric structure on the manifold of horospheres that guarantees the possibility of applying this formula.

This lecture is correspondingly divided into three parts: obtaining a general inversion formula in curvilinear problems of integral geometry; verifying the applicability of this formula in complex semisimple Lie groups; and, finally, describing the geometric structure on the manifold of horospheres.

1. Integral-geometric introduction

1.1. The Gel'fand-Graev-Shapiro operator $\varkappa$ in the plane problem of integral geometry [5]. Let us consider the manifold $H = H_{n,p}$ of p-dimensional planes in $\mathbb{C}^n_z$ with fixed parametrization. Let

$$\pi(\alpha,\beta), \qquad \alpha = (\alpha^1, \ldots, \alpha^p), \qquad \alpha^j, \beta \in \mathbb{C}^n,$$

have the form

$$z = \alpha t + \beta = \sum_{j=1}^{p} \alpha^j t_j + \beta, \qquad t = (t_1, \ldots, t_p) \in \mathbb{C}^p; \qquad (1)$$

where (α, β) are coordinates on H. We associate with $f(z) \in C_0^\infty(\mathbb{C}^n)$ its integrals over $\pi(\alpha, \beta)$:

$$\hat{f}(\alpha, \beta) = \int_{\mathbb{C}^p} f(\alpha t + \beta) dt \wedge \overline{dt}, \qquad dt = dt_1 \wedge \cdots \wedge dt_p. \qquad (2)$$

We shall denote by H_z the submanifold of planes $\pi(a, z)$ (the pass through the point z). We consider the operator $\varkappa_j$ on forms that raise the degree of the form by 1:

$$\varkappa_j = \frac{\partial}{\partial \beta} \cdot d\alpha^j = \sum_i \frac{\partial}{\partial \beta_i} d\alpha^j_i, \qquad \overline{\varkappa}_j = \frac{\partial}{\partial \overline{\beta}} \cdot \overline{d\alpha}^j. \qquad (3)$$

Let

$$\varkappa = \varkappa_1 \wedge \cdots \wedge \varkappa_p, \qquad \overline{\varkappa} = \bigwedge \overline{\varkappa}_j.$$

PROPOSITION [5]. *The (p,p)-form $(\varkappa \wedge \overline{\varkappa})\hat{f}|_{H_z}$ for each $z \in \mathbb{C}^n$ is closed on H_z, and if γ is a $2p$-dimensional (over $\mathbb{R}$) cycle in H_z, then*

$$\int_\gamma (\varkappa \wedge \overline{\varkappa})\hat{f} = c(\gamma)f(z), \qquad (4)$$

where $c(\gamma)$ does not depend on f.

Only the fact that $(\varkappa \wedge \overline{\varkappa})\hat{f}|_{H_z}$ is closed is needed in the proof, and this can be verified immediately.

We note that if for some function φ on H the forms $(\varkappa \wedge \overline{\varkappa})\varphi|_{H_z}$ are closed, then $\varphi = \hat{f}$ for some function f.

Let us discuss what formula (4) yields for integral geometry. Let K be a submanifold in H of (complex) dimension n. We wish to reconstruct $f(z)$ in terms of $\hat{f}|_K$. Let $K_z = K \cap H_z$. We can assume that $\dim_\mathbb{C} K_z = p$ for almost all z. If the K_z are cycles, where $c(K_z) \neq 0$, then, using (4), we can reconstruct $f(z)$. However, there is one very delicate moment in this situation: to compute $(\varkappa \wedge \overline{\varkappa})\hat{f}|_{H_z}$ it may not be enough to know $\hat{f}$ only on K (there may not be enough differentiations in tangent directions). Only for very special submanifolds K is this possible (they are called admissible). However, in particular, it turned out that the manifold of horospheres in $SL(l, \mathbb{C})$ is admissible, which made it possible to obtain the Plancherel formula for this group in [6].

It is important to know how to calculate the coefficient $c(\gamma)$. This can be done by means of test functions, but it is useful to take into account its geometric significance: $c(\gamma)/(2\pi)^{2p}$ is the number of planes $\pi \in \gamma$ contained in a general hyperplane.

1.2. The closed extension of $\varkappa$ to curvilinear submanifolds.

The main point of the preceding discussion was that the form $(\varkappa \wedge \overline{\varkappa})\hat{f}$ is closed. We now wish to extend it, keeping it closed, to the manifold (infinite-dimensional!) of all p-dimensional submanifolds. Our considerations, as before, are local. Let Π be the set of *all* smooth parametrized submanifolds in a neighborhood of zero $U \subset \mathbb{C}^n_z$:

$$z = \varphi(t), \qquad \varphi = (\varphi_1, \ldots, \varphi_n), \qquad z \in \mathbb{C}^n, \qquad t \in \mathbb{C}^p. \qquad (5)$$

For $f(z) \in C_0^\infty(U)$ we set

$$\hat{f}(\varphi) = \int f(\varphi(t))\, dt \wedge \overline{dt}. \qquad (6)$$

We denote by Π_z the set of φ for which $\varphi(0) = z$. We have $H \subset \Pi$. For $\pi(\alpha, \beta)$ we set $\varphi(t) = \alpha t + \beta$. For simplicity we shall carry out the calculations for $z = 0$.

We shall identify elements of the tangent space $T_\varphi \Pi$ with the variations $\delta\varphi$. We introduce in $T_\varphi \Pi$ the canonical decomposition into subspaces

$$T_\varphi \Pi_c = T^{(1)} \oplus \cdots \oplus T^{(p)} \qquad (7)$$

and we shall denote by $\delta^{(j)}\varphi$ the components of $\delta\varphi$ in $T^{(j)}$; the following conditions must hold:

(i) $\delta^{(j)}\varphi(t)$ depends only on $t_1, \ldots, t_j$;

(ii) $\delta^{(j)}\varphi|_{t_j=0} \equiv 0$.

By virtue of these conditions,

$$\delta^{(j)}\varphi(t) = \delta\varphi(t_1, \ldots, t_j, 0, \ldots, 0) - \delta\varphi(t_1, \ldots, t_{j-1}, 0, 0, \ldots, 0),$$

and the decomposition (7) is uniquely defined.

On the functionals $F(\varphi)$ on Π we define the following operator in 1-forms on Π_0:

$$\varkappa_j F(\varphi; \delta_\varphi) = \delta F(\varphi; \delta^{(j)}\varphi/t_j) \tag{8}$$

and we extend it to forms. In other words, it is necessary to take the value of the variation δF of the functional F a the point φ in the variation $\delta^{(j)}\varphi/t_j$. By (ii), this is a regular variation, but, of course, it is not tangent to Π_0. In particular, for the functional $\hat{f}(\varphi)$ we have

$$\varkappa_j \hat{f} = \int (\langle \operatorname{grad} f(\varphi(t)), \delta^j \varphi(t) \rangle / t_j) \, dt \wedge \overline{dt}.$$

Setting $\varkappa = \bigwedge \varkappa_j$, we consider the (p, p)-form

$$(\varkappa \wedge \overline{\varkappa})\hat{f} = \int \sum \frac{\partial^{2p} f(\varphi(t))}{\partial z_{i_1} \cdots \partial z_{i_p} \partial \bar{z}_{j_1} \cdots \partial \bar{z}_{j_p}} \delta^{(1)}\varphi_{i_1} \wedge \cdots \wedge \delta^{(p)}\varphi_{j_p} \wedge \frac{dt \wedge \overline{dt}}{\Pi t_{i_k} \overline{\Pi t}_{j_l}}.$$

Two basic properties of the form $(\varkappa \wedge \overline{\varkappa})\hat{f}$ can be verified immediately:

(A) This is a closed form on Π_0.

(B) On the submanifold of planes $H_0 \subset \Pi_0$ it coincides with the form in §1.1 (see (3)).

The forms $(\varkappa \wedge \overline{\varkappa})F|_{\Pi_z}$ can be defined in a similar manner. The closedness of all these forms is equivalent to the fact that $F = \hat{f}$ for some function f.

THEOREM. *For any $2p$-dimensional (over $\mathbb{R}$) cycle $\gamma \subset \Pi_z$ we have*

$$\int_\gamma (\varkappa \wedge \overline{\varkappa})\hat{f} = c(\gamma) f(z). \tag{9}$$

This follows from (A), (B), and the proposition in §1.1. As regards the calculation of $c(\gamma)$, it is useful to keep in mind that $c(\gamma) = c(\tilde{\gamma})$, where γ is the cycle of tangent planes to the submanifolds $\varphi \in \gamma$ at the point z (the cycles γ and $\tilde{\gamma}$ are homologous).

Everything said in the previous subsection about the connection with integral geometry carries over to the more general situation under consideration here: with the help of (9) one can solve the problem of integral geometry for the manifolds $K \subset \Pi$, $\dim_{\mathbb{C}} K = n$, for which the forms $(\varkappa \wedge \overline{\varkappa})F|_{K_z}$ for almost all z are determined by $F|_K$, the K_z are cycles and $c(K_z) \neq 0$. We shall see that manifolds of horospheres in complex semisimple Lie groups have this property.

2. Solution of the integral-geometry problem in complex semisimple Lie groups

2.1. Notation. Let G be such a group, $\dim G = n$, $\mathfrak{g}$ its Lie algebra, H and $\mathfrak{h}$ the corresponding Cartan subgroup and subalgebra, $\dim \mathfrak{h} = l$, and $\{\alpha\}$ a system of positive roots. We index these roots so that

$$\text{if } \alpha_i + \alpha_j = \alpha_k, \quad \text{then} \quad k < i, \quad k > j. \tag{10}$$

The existence of such an indexing (nonunique) can be proved easily [10]. Let $e_{\pm j}$, $j > 0$, denote the root vectors for $\pm \alpha_j$ respectively, and let

$$f_e = [e_{-i}, e_i], \qquad [f_i, e_j] = \langle f_i, f_j \rangle e_j,$$

where $\langle \, , \, \rangle$ is the Cartan scalar product. As a basis in $\mathfrak{g}$ we choose all $\{e_{\pm i}\}$ and f_i for simple roots. We introduce exponential coordinates in G (we confine ourselves to local discussions in a neighborhood of the identity $e \in G$).

Let $\mathfrak{z}$ be the subalgebra spanned by e_i, $i > 0$, $\dim \mathfrak{z} = (n-l)/2 \stackrel{\text{def}}{=} p$, Z the corresponding maximal unipotent subgroup: $Z = \{\exp(\sum_{j>0} t_j e_j)\}$, and Z_- the subgroup corresponding to the negative roots. The horospheres in G have the form $Z(g_1, g_2) = g_1 Z g_2$. Almost all horospheres are obtained if we confine ourselves to $g_1 = \zeta_1 h$, $g_2 = \zeta_2$, $\zeta_1, \zeta_2 \in Z$, $h \in H$. Correspondingly, $\{\zeta_1, \zeta_2, h\}$ define coordinates in an everywhere dense chart on the manifold of horospheres Ξ, $\dim \Xi = n$. Almost all horospheres that pass through e (Ξ_e is their totality) have the form

$$Z(\zeta^{-1}, \zeta), \qquad \zeta \in Z_-, \qquad \dim \Xi_e = \dim Z = p.$$

For $f \in C_0^\infty(G)$ we consider the integrals over horospheres

$$(\zeta_1, \zeta_2, h) = j(h) \int f\left(\zeta_1 h \exp\left(\sum_{j>0} t_j e_j\right) \zeta_2\right) dt \wedge \overline{dt}, \tag{11}$$

where

$$j(\exp f) = \exp\left(\frac{1}{2}\left\langle f, \sum_{i>0} f_i \right\rangle\right), \qquad f \in \mathfrak{h}.$$

2.2. Calculation of the restriction of the form $(\varkappa \wedge \overline{\varkappa})\hat{f}$ to the manifold of horospheres. From homogeneity considerations it is enough to carry out the calculations at the identity $e \in G$, that is, to calculate $(\varkappa \wedge \overline{\varkappa})\hat{f}|_{\Xi_e}$. This is a form on Z_- by virtue of the indicated parametrization of Ξ_e. But again from homogeneity considerations it is enough to carry out the calculations at $\zeta = e$ on Z_-.

Thus, it is necessary to investigate the variation δZ of the horosphere Z in Ξ_e. We consider the following exponential coordinates on Z_- : $\zeta = \exp(\sum_{i>0} s_i e_{-i})$. Then

$$\delta Z(t) = d_s \ln \left[\zeta^{-1} \exp\left(\sum t_j e_j \right) \zeta \right],$$

where we take the differential with respect to s for $s = 0$. Of course, δZ takes values in the Lie algebra $\mathfrak{g}$. Using the formula

$$\exp(-\varepsilon Y) \exp(X) \exp(\varepsilon Y) = \exp(X + \varepsilon [X, Y] + o(\varepsilon)),$$

we obtain

$$\delta Z = \left[\sum_{j>0} t_j e_j, \sum_{l>0} ds_l e_{-l} \right],$$

and for the components of the canonical decomposition we have

$$\delta^{(j)} Z = t_j \left[e_j, \sum_{l>0} ds_l e_{-l} \right]. \qquad (12)$$

As must be the case, $\delta^{(j)} Z / t_j$ is a regular variation, but in our case, moreover, it does not depend on t. In addition, these variations will not be tangent to the manifold of horospheres. Nevertheless, we shall see that $(\varkappa \wedge \overline{\varkappa}) \hat{f}$ can be computed. Let us carefully investigate $\delta^{(j)} Z / t_j$:

$$\delta^{(j)} Z / t_j = -ds_j f_j + \delta_1^{(j)} Z + \delta_2^{(j)} Z,$$

where
 (i) $\delta_1^{(j)} Z$ decomposes with respect to ds_k, $k > j$;
 (ii) $\delta_2^{(j)} Z$ takes values in the Lie algebra $\mathfrak{z}$ of the group Z and decomposes with respect to e_m, $m < j$.

We obtain the first term if we take $l = j$ in (12). In $\delta_1^{(j)} Z$ we group the terms with $l > j$, and in $\delta_2^{(j)} Z$ we group the terms with $l < j$. It is necessary to verify only that (ii) is valid.

Let us investigate $[e_j, e_{-l}]$. It is necessary to verify that if $\alpha_j - \alpha_l$ is a root, then it is a positive root α_m. Indeed, let $\alpha_j - \alpha_l = -\alpha_k$. Then by property (10) of the root indexing $j > l$ and we arrive at a contradiction. But if $\alpha_j - \alpha_l = \alpha_m$, then again by (10) we have $m < j$.

Let us calculate $\varkappa \hat{f}|_{\Xi_e}$ for $\zeta = e$. We recall that to calculate $\varkappa_j$ we must find the value of the variation on $\delta(j) Z_1 t_j$. Let us first note that the $\delta_2^{(j)} Z$ contribute nothing since, by (ii), they correspond to translations along the horosphere Z itself that preserve an element of volume dt and, hence, the integral $\hat{f}$.

We shall show by induction on decreasing j that the $\delta_1^{(j)} Z$ contribute nothing as well. For $j = p = (n-l)/2$ we have $\delta_1^{(j)} Z = 0$. Furthermore, by induction we see that $\bigwedge_{j>k} \varkappa_j \hat{f}$ has the form $c \bigwedge_{j>k} ds_j$. Indeed, with the addition of $\varkappa_k$, $\delta_1^{(k)} Z$ contributes nothing, since by (i) it is a combination of the ds_m, $m > j$. Hence $\bigwedge_{j \geq k} \varkappa_j \hat{f}$ has the form $\tilde{c} \bigwedge_{j \geq k} ds_j$. Therefore, for each j only the value of the variation on $-ds_j f_j$ is essential, but this is the tangent variation to Ξ corresponding to translations by elements of a Cartan subgroup H. As a result, if the derivation in the direction of f_j is denoted by $\mathscr{D}_j$, then it has been proved that

$$\varkappa \hat{f}(e, e, e) = (-1)^p \prod_{j>0} \mathscr{D}_j \hat{f}(e, e, h)|_{h=l} \bigwedge_{i>0} ds_i.$$

Here is it necessary to bear in mind that

$$\hat{f}(e, e, \exp(\varepsilon f)) = \int f\left(\exp\left(\varepsilon f + \sum t_j e_j\right)\right) dt \wedge dt + o(\varepsilon).$$

For the sake of this relation the normalization factor $j(h)$ was introduced in definition (11). The operator $\bar{\varkappa}$ has a similar form, and the same formulas by homogeneity carry over to other points $\zeta \in Z_-$. As a result,

$$(\varkappa \wedge \bar{\varkappa}) \hat{f}(\zeta^{-1}, \zeta, e) = \prod_{j>0} \mathscr{D}_j \overline{\mathscr{D}}_j \hat{f}(\zeta^{-1}, \zeta, h)|_{h=l} \bigwedge ds_j \wedge \overline{ds}_j.$$

It remains to examine the integral $\int_{\Xi_e} (\varkappa \wedge \bar{\varkappa}) \hat{f}$ and to calculate $c(\Xi_e)$.

For a cycle Ξ_e the cycle of tangent planes $\widetilde{\Xi}_e$ consists of subspaces in $\mathfrak{g}$ of the form $Tg\mathfrak{z}$, where $g \in G$ and Tg is the adjoint representation of G in $\mathfrak{g}$. It is enough to take $g \in U$, where U is a maximal compact subgroup and almost all planes are taken for $g \in Z_-$. Let π_u be the hyperplane in $\mathfrak{g}$ of the form $\langle u, x \rangle = 0$, $u \in \mathfrak{g}$. We can confine ourselves to the case $u \in \mathfrak{h}$, where regular elements of $\mathfrak{h}$ correspond to general hyperplanes. In this case π_u contains $|W|$ planes in $\widetilde{\Xi}_e$, where $|W|$ is the order of the Weyl group (algebras $\mathfrak{z}$ corresponding to different orderings of roots in $\mathfrak{h}$). It is easy to deduce this, for example, from the Bruhat decomposition. Thus, $c(\Xi_e) = (2\pi)^{n-l}|W|$ and we have

$$f(e) = \frac{1}{(2\pi)^{n-l}|W|} \int_{Z_-} \prod_{j>0} \mathscr{D}_j \overline{\mathscr{D}}_j \hat{f}(\zeta^{-1}, \zeta, h)|_{h=l} \bigwedge_{j>0} ds_j \wedge d\bar{s}_j. \quad (13)$$

3. Infinitesimal structure on the manifold of horospheres

In conclusion, we shall present without proof the geometric structure on the manifold of horospheres that is sufficient for the existence of an inversion formula based on the form $(\varkappa \wedge \bar{\varkappa}) \hat{f}$.

Let us investigate the picture of the incidence of horospheres of the projective duality type. We consider the manifold of horospheres Ξ, $\dim \Xi = n$. To group elements $g \in G$ there correspond on Ξ submanifolds Ξ_g of horospheres that pass through g, $\dim \Xi_g = p = (n-l)/2$. Let $\xi \in \Xi$ (a horosphere $Z(\xi)$ in G is fixed). We consider submanifolds $\Xi_g \ni \xi$ (that is,

$g \in Z(\xi)$) and the tangent planes $\sigma_g \subset T_\xi \Xi$ to Ξ_g at the point ξ. A p-parameter family of p-planes arises in the tangent space $T_\xi \Xi$.

It turns out that the configuration of these planes has a remarkable property which largely determines the inversion formula. By the *p-sheaf* (σ, λ) we mean the union of p-planes that lie in a fixed $(p+1)$-plane and contain a fixed $(p-1)$-plane; σ is the *axis of the sheaf*.

Then the family of planes σ_g splits into a $(p-1)$-parameter family of p-sheaves. The axes of these sheaves $\{\sigma^{(p-1)}\}$, in turn, split into a $(p-2)$-parameter family of $(p-1)$-sheaves, etc. At the final stage we find that the one-parameter family of lines that are the axes of the 2-sheaves at the preceding stage lie in one plane and pass through the origin. This stratification corresponds to the root numbering scheme (10) and, to the same extent, is nonunique: at each stage a finite number of variants of the partitioning into sheaves can be encountered. The axes of the sheaves can be linked with degenerate horospheres.

If for some family Ξ of p-dimensional submanifolds on an n-dimensional manifold G the indicated inductive partitioning into sheaves of tangent planes holds at each point $\xi \in \Xi$, then we say that Ξ satisfies condition $(\mathscr{H})$, or the *horospherical* condition. For a solution of the problem of integral geometry to be possible, one more condition must hold.

We say that Ξ satisfies the *infinitesimal Desarguesian condition* if in a neighborhood of each point $\xi \in \Xi$ there exists a diffeomorphism that rectifies Ξ_g to within a magnitude of the 3rd order of smallness. This condition can be expressed analytically.

In the family Ξ of submanifolds $Z_\zeta \subset G$ that satisfy condition $(\mathscr{H})$ and the infinitesimal Desarguesian condition the form $(\varkappa \wedge \overline{\varkappa})\hat{f}$ induces an inversion formula.

Condition $(\mathscr{H})$ permits a "G-representation" reformulation: on the submanifold Z_ξ (in particular, on horospheres) there is a canonical triangular-rational structure; namely, there are a mapping of Z_ξ onto $\mathbb{C}p^1$, a mapping of the inverse images of points again onto $\mathbb{C}p^1$, etc. This is the geometric expression of the root structure, and it certainly merits careful deliberation. In particular, it would be interesting to explain how these considerations are related to the integration of nonlinear equations representable as a compatibility condition for a system of linear equations with several spectral parameters.

Bibliography

1. I. M. Gel′fand and M. A. Naĭmark, *Unitary representations of the classical groups*, Trudy Mat. Inst. Steklov, Vol. 36, Akad. Nauk SSSR, Moscow and Leningrad, 1950; German transl., Akademie-Verlag, Berlin, 1957.

2. Harish-Chandra, *Plancherel formula for complex semisimple Lie groups*, Proc. Nat. Acad. Sci. **37** (1951), 813.

3. I. M. Gel′fand and M. I. Graev, *Analog of the Plancherel formula for the classical groups*, Trudy Moskov. Mat. Obshch., Vol. 4, Moscow, 1955, pp. 375–404; English transl. in Amer. Math. Soc. Transl. (2) **9** (1958), 123–154.

4. ____, *Geometry of homogeneous spaces, representations of groups in homogeneous spaces, and problems of integral geometry connected with them*, Trudy Moskov. Mat. Obshch., Vol. 8, Moscow, 1959, pp. 321–390.

5. I. M. Gel′fand, M. I. Graev, and Z. Ya. Shapiro, *Integral geometry on k-dimensional planes*, Funktsional Anal. i Prilozhen. **1** (1967), 15–31; English transl. in Functional Anal. Appl. **1** (1967).

6. I. M. Gel′fand and M. I. Graev, *Complexes of k-dimensional planes in the space $\mathbb{C}^n$ and Plancherel's formula for the group $GL(n, \mathbb{C})$*, Dokl. Akad. Nauk SSSR **179** (1968), 522–525; English transl. in Soviet Math. Dokl. **9** (1968), 394–398.

7. I. M. Gel′fand, S. G. Gindikin, and Z. Ya. Shapiro, *A local integral geometry problem in spaces of curves*, Funktsional. Anal. i Prilozhen. **13** (1979), 11–31; English transl. in Functional Anal. Appl. **13** (1979).

8. S. G. Gindikin, *Reductions of varieties of rational curves and related problems in the theory of differential equations*, Funktsional. Anal. i Prilozhen. **18** (1984), 14–39; English transl. in Functional Anal. Appl. **18** (1984).

9. R. Penrose, *Nonlinear gravitons and curved twistor theory*, Gen. Relativity Gravitation **7** (1976), 31–52.

10. S. G. Gindikin and F. I. Karpelevich, *Plancherel measure for Riemann symmetric spaces of nonpositive curvature*, Dokl. Akad. Nauk SSSR **145** (1962), 252–255; English transl. in Soviet Math. Dokl. **3** (1962), 962–965.

Translated by R. LENET

Superalgebras and Identities

E. I. ZEL'MANOV

It is known [1] that many interesting phenomena in infinite-dimensional algebras manifest themselves already in the case of finite-dimensional algebras over fields of prime characteristics. In this survey we will try to show how the properties of a variety depend on the structure of its infinite-dimensional superalgebras.

In §1 we discuss the structure of associative, alternative and Jordan superalgebras. As for the most substantive theory of Lie superalgebras, it is beautifully presented in the surveys [2], [3] and the monograph [4], and we find it hard to add anything to it.

In §2 we show how superalgebras can be used to derive certain identities from others and to study properties of infinite-dimensional algebras.

§1. Superalgebras

As usual, a superalgebra is a Z_2-graded algebra $A = A_0 + A_1$, $A_i A_j \subseteq A_{i+j}$ (modulo 2). An example of such an object is the Grassmann algebra G given by a set of generators $1, e_1, e_2, \ldots$ together with defining relations $e_i^2 = 0$, $e_i e_j + e_j e_i = 0$, where $i \neq j$. The products $1, e_{i_1} \cdots e_{i_k}$, $i_1 < \ldots < i_k$, form a basis of the algebra G; the products of even length (even number of factors) form a basis of the even component G_0, and the products of odd length form a basis of the odd component G_1, $G = G_0 + G_1$.

Given an arbitrary superalgebra $A = A_0 + A_1$, we consider the algebra tensor product $A \otimes G$. Its subalgebra $A_0 \otimes G_0 + A_1 \otimes G_1$ is called the Grassmann envelope of A and it is denoted by $G(A)$.

DEFINITION. Let $\mathfrak{M}$ be a variety of linear algebras (in the sense of [5], [6]). We say that a superalgebra $A = A_0 + A_1$ is an $\mathfrak{M}$-superalgebra if $G(A) \in \mathfrak{M}$.

It can be shown that $G(A)$ satisfies the identities defining $\mathfrak{M}$ if and only if A satisfies a certain system of graded identities.

1980 *Mathematics Subject Classification* (1985 *Revision*). Primary 17A70; Secondary 17A65.

1.1. Associative superalgebras. A superalgebra $A = A_0 + A_1$ is an associative superalgebra if and only if it is associative in the usual sense. In addition, there is an involutive automorphism of parity on A that transforms a homogeneous element $a_i \in A_i$ into $(-1)^i a_i$. Using this remark it is easy to verify (see [7]) that any simple finite-dimensional associative superalgebra over an algebraically closed field is isomorphic to one of the following superalgebras:

1) $A = A_0 = M_n(\Phi)$, the algebra of n by n matrices;

2) $A = \{\begin{pmatrix} A & B \\ B & A \end{pmatrix} | A, B \in M_n(\Phi)\}$, $A_0 = \{\begin{pmatrix} A & 0 \\ 0 & A \end{pmatrix}\}$, $A_1 = \{\begin{pmatrix} 0 & B \\ B & 0 \end{pmatrix}\}$;

3) $A = M_{p,q}(\Phi) = M_{p+q}(\Phi)$, $A_0 = \{\begin{pmatrix} * & 0 \\ 0 & * \end{pmatrix}\}$, $A_1 = \{\begin{pmatrix} 0 & * \\ * & 0 \end{pmatrix}_q^p\}$.

1.2. Alternative algebras and superalgebras. An algebra is called alternative if it satisfies the following two identities:

(A1) $$x^2 \cdot y = x(xy),$$
(A2) $$y \cdot x^2 = (yx)x.$$

The validity of (1) and (2) in an algebra A is equivalent to the fact that all two-generator superalgebras in A are associative (E. Artin).

Basic examples of such algebras are associative algebras and octonions (see [8]).

We recall that an algebra is called prime if the product of any two nonzero ideals in it is itself nonzero.

If the characteristic of the ground field is not zero, then, essentially, there are no nontrivial alternative algebras.

THEOREM 1.1 [9]. *Any prime alternative superalgebra $A = A_0 + A_1$ over a field of characteristic $\neq 3$ is either associative or else $A_1 = (0)$ and A_0 is the octonion algebra.*

1.3. Jordan algebras and superalgebras. Jordan algebras were first introduced in a paper of P. Jordan, J. von Neumann, and E. Wigner [10]. In the classical interpretation of the quantum mechanics the observables are the Hermitian operators in a Hilbert space. The set of all Hermitian operators is closed under linear combinations $\alpha x + \beta y$ (with real coefficients), but it is not closed under composition xy. Still, it is closed with respect to the symmetrized product $x \circ y = (1/2)(xy + yx)$. The authors' point was (1) to formulate some important algebraic properties of Hermitian operators in terms of the operation $\circ$; (2) to study all algebraic systems possessing these properties. The following identities were chosen by the authors as basic properties:

(J1) $$x \circ y = y \circ x \quad \text{(commutativity)};$$
(J2) $$x^2 \circ (y \circ x) = (x^2 \circ y) \circ x \quad \text{(Jordan identity)}.$$

At present a linear algebra over a field Φ (of characteristic $\neq 2$) satisfying (J1) and (J2) is called a (linear) Jordan algebra.

Now we consider the most important examples of Jordan algebras: (1) if we replace the multiplication in an associative algebra R by the symmetrized product $x \circ y = (1/2)(xy+yx)$, then R becomes a Jordan algebra, denoted by $R^{(+)}$; (2) a linear operator $*\colon R \to R$ is called an involution if $(a^*)^* = a$, $(ab)^* = b^* a^*$ for any $a, b \in R$. The set of $*$-symmetric elements $H(R, *) = \{a \in R | a^* = a\}$ is a subalgebra in $R^{(+)}$; (3) let $f\colon V \times V \to \Phi$ be a symmetric bilinear form in a vector Φ-space V. We introduce on the direct sum $\Phi \cdot 1 + V$ a multiplication by setting $v \circ w = f(v, w) \cdot 1$, $v, w \in V$. It makes $\Phi + V$ a Jordan algebra; (4) let $\mathbf{O}$ be the octonion algebra over a field Φ with canonical involution $^-\colon \mathbf{O} \to \mathbf{O}$. We consider 3 by 3 matrices over $\mathbf{O}$ with involution $*\colon \mathbf{O}_3 \ni (x_{ij}) \to (\bar{x}_{ji})$. Then $H(\mathbf{O}_3) = \{A \in \mathbf{O}_3 | A^* = A\}$ is a Jordan algebra with respect to the symmetrized product of matrices. It is known that $H(\mathbf{O}_3)$ is simple and $\dim_\Phi H(\mathbf{O}_3) = 27$.

A Jordan algebra is called special if it can be embedded in $R^{(+)}$, where R is an associative algebra. Otherwise we say that the Jordan algebra is exceptional. It is easy to see that the algebras in (1), (2), and (3) are special. A. Albert has shown in [11] that algebras of type (4) are exceptional.

A curious fact is that, essentially, the quantum mechanics deals with Jordan superalgebra rather than with Jordan algebra. Indeed, the dynamics of the observables is given by the Poisson bracket

$$\{X, Y\} = \frac{i}{\hbar}[X, Y],$$

where X, Y are the Hermitian operators, $i^2 = -1$, and $\hbar$ is the Plank constant.

We identify both the even and the odd spaces J_0 and J_1 with the Jordan algebra of Hermitian operators H and let J_0 act on J_1 by the symmetrized product. Then the Poisson bracket $\{,\}$ on J_1 turns $J_0 + J_1$ into a Jordan superalgebra.

If in an arbitrary associative superalgebra $R = R_0 + R_1$, we replace the product in R_0 and the action of R_0 on R_1 by the symmetrized product and the product of elements in R_1 by the commutator, then we obtain a Jordan superalgebra $R_0^{(+)} + R_1^{(-)}$. A graded linear map $*\colon R_0 + R_1 \to R_0 + R_1$ is called a superinvolution if $(a_i b_j)^* = (-1)^{ij} b_j^* a_i^*$, $(a_i^*)^* = a_i$, where $a_i \in R_i$, $b_j \in R_j$. Then the superspace of the elements fixed by $*$, $H(R, *) = H(R_0, *) + H(R_1, *)$, is a Jordan superalgebra. It is in this way that we get the Jordan algebra of quantum mechanics. Since we know all simple finite-dimensional associative superalgebras, it is easy to describe up to equivalence all their superinvolutions (see [3]).

Now suppose we are given a non-degenerate bilinear form Q on a Φ-superspace $V = V_0 + V_1$ such that Q is symmetric on V_0 and alternating on V_1, with $Q(V_0, V_1) = Q(V_1, V_0) = 0$. Then the product of the form

$v \circ w = Q(v, w) \cdot 1$ makes $(1 \cdot \Phi + V_0) + V_1$ a Jordan superalgebra $J(Q)$.

The following Jordan superalgebras are given in [12]:

3-dimensional superalgebra $K = J_0 + J_1$, $J_0 = \Phi e$, $e^2 = e$, $J_1 = \Phi y_1 + \Phi y_2$, $e \circ y_i = (1/2) y_i$, $y_1 \circ y_2 = e$;

the family of 4-dimensional superalgebras D_τ ($\tau \neq 0$, $\tau \in \Phi$), $J_0 = \Phi e_1 \oplus \Phi e_2$ is the direct sum of fields, $J_1 = \Phi y_1 + \Phi y_2$, $e_i \circ y_j = (1/2) y_j$, $y_1 \circ y_2 = \tau e_1 + e_2$;

10-dimensional exceptional Jordan superalgebra F_{10} associated by I. L. Kantor's construction with 40-dimensional exceptional Lie superalgebra.

It was noticed by I. L. Kantor that an interesting series of Hamiltonian superalgebras $H(n)$ is missing in [12], the superalgebras in this series being associated by I. L. Kantor's construction with Hamiltonian Lie superalgebras.

THEOREM 1.2 (V. G. Kac, I. L. Kantor). *Any simple finite-dimensional Jordan superalgebra over an algebraically closed field of characteristic zero is isomorphic to one of the superalgebras in the list*

$$R_0^{(+)} + R_1^{(-)}, \ H(R_0 + R_1, *), \ J(Q), \ K, \ D_\tau, \ F_{10}, \ H(n), \ H(\mathbf{O}_3).$$

A Jordan superalgebra is called special if it can be embedded in a superalgebra of the form $R_0^{(+)} + R_1^{(-)}$. Otherwise we say that the superalgebra is exceptional.

All the above-mentioned superalgebras except F_{10}, $H(n)$ (and, of course, $H(\mathbf{O}_3)$) are special. It can be shown that F_{10} and $H(n)$ are, indeed, exceptional.

Finally, we consider an interesting example of an infinite-dimensional Jordan superalgebra of vector fields. Its even and odd components J_0 and J_1 are identified with the polynomial algebra $\Phi[x]$ over a field of characteristic zero, the product in J_0 and the action of J_0 on J_1 being just multiplication by polynomials, the bracket $[f(x), g(x)] = f'(x)g(x) - f(x)g'(x)$ being the Poisson bracket.

This superalgebra has finite-dimensional analogies over prime characteristic fields. Namely, let $O(x)$ be the algebra of truncated polynomials over a field Φ of characteristic $p > 0$, i.e., $O(x) = \Phi[x]/(x^p = 0)$ and let the derivation transform x into 1. Then $J_0 = J_1 = O(x)$, $[f, g] = f'g - fg'$, is a simple finite-dimensional Jordan superalgebra over Φ.

§2. Identities

In this section we show how one can use superalgebras to derive certain identities from others, when a direct deduction is unlikely. This approach was first used, most probably, by Kemer in [13].

In what follows all algebras are considered over a field Φ of characteristic zero.

Let A be a linear algebra, and f a noncommutative polynomial. Denote by $V_f(A)$ the ideal generated in A by the set of values $f(A)$. We will be interested in the following problem.

GLOBAL NILPOTENCE PROBLEM. When is $V_f(A)$ a nilpotent ideal?

Many of the problems discussed in the literature can be reduced to this (see below). Let us explain the idea of the approach via superalgebras.

Let $\mathfrak{M}$ be the variety of linear Φ-algebras, $A_{\mathfrak{M}}$ its free algebra with the set of generators $X = \{x_i, i \geq 1\}$; $\Lambda = \{x_{i_1}, \ldots, x_{i_m}\} \subset X$, $i_1 < \cdots < i_m$. An arbitrary permutation $\sigma \in S_m$ induces an automorphism $\hat{\sigma}$ of $A_{\mathfrak{M}}$: $x_{i_k} \to x_{i_{\sigma(k)}}$, $x_j \to x_j$ for $x_j \notin \Lambda$. Let $H_\Lambda = \sum_{\sigma \in S_m} \hat{\sigma}$ denote the symmetrization over Λ, and $S_\Lambda = \sum_{\sigma \in S_m} (-1)^{|\sigma|} \hat{\sigma}$ the skew-symmetrization over Λ. The following is an obvious consequence of the Young diagram method (see [14]).

LEMMA. *Let $h(\ldots, x_1, \ldots, x_m, \ldots)$ be a nonzero element in $A_{\mathfrak{M}}$, multilinear in $x_1, \ldots, x_m$. Then there exits a subset $\Lambda = \{x_{i_1}, \ldots, x_{i_t}\} \subset \{x_1, \ldots, x_m\}$, $t \geq [\sqrt{m}]$, and a permutation $\tau \in S_m$ such that either $H_\Lambda h^\tau \neq 0$ or $S_\Lambda h^\tau \neq 0$.*

Now if an expression $h(\ldots, c_1, \ldots, c_k, \ldots)$ is multilinear and symmetric in $c_1, \ldots, c_k$, then we can replace all $c_1, \ldots, c_k$ by a unique new free generator. Obviously we obtain a nonzero expression.

If h is skew-symmetric in $c_1, \ldots, c_k$, however, then one can use the following device remarkable both by its simplicity and ingenuity.

KEMER'S SUPERTRICK. We replace each of the elements $c_1, \ldots, c_k$ by the sum $c = \sum_{i=1}^{k} c_i \otimes e_i$ where $e_1, \ldots, e_k$ are the Grassmann variables in the Grassmann algebra G. Then

$$h(\ldots, \sum c_i \otimes e_i, \ldots, \sum c_i \otimes e_i, \ldots)$$
$$= (-1)^{|\sigma|} h(\ldots, c_{\sigma(1)}, \ldots, c_{\sigma(k)}, \ldots) \otimes e_1 \cdots e_k$$
$$= k! h(\ldots, c_1, \ldots, c_k, \ldots) \otimes e_1 \cdots e_k \neq 0.$$

In this way we arrive at a superalgebra of the form $A \otimes G$.

Using supertrick we can reduce, in a number of cases, the global nilpotence problem to the following.

PRIME SUPERALGEBRA PROBLEM. Let $A = A_0 + A_1$ be a prime $\mathfrak{M}$-superalgebra. Is it true that $f = 0$ holds identically in the Grassmann envelope $G(A)$?

As an example we consider the description, due to Kemer, of T-prime varieties of associative Φ-algebras.

A variety $\mathfrak{M}$ is called T-prime if for any two noncommutative polynomials f, g the equality $V_f(A_{\mathfrak{M}}) V_g(A_{\mathfrak{M}}) = 0$ implies that either $f(A_{\mathfrak{M}}) = 0$ or $g(A_{\mathfrak{M}}) = 0$.

THEOREM 2.1 [15]. *Any T-prime variety of associative Φ-algebras is generated by the Grassmann envelope of a simple finite-dimensional associative Φ-superalgebra.*

Considering the classification of simple finite-dimensional associative superalgebras we see that there exist three series of such varieties corresponding to three series of superalgebras.

Now let Alt[X] be the free alternative algebra in countably many free generators. Shestakov [16] and Slater [17] have shown that its radical consists of elements which are identically zero in all associative algebras and in the octonion algebra. If it were true, as people sometimes say, that the theory of alternative algebras studies and unites associative algebras and octonions, then these identities (from the radical) should be incorporated into the definition of alternative algebras. Then, of course, the definition would become overcomplicated.... Thus, the radical $J(\text{Alt}[X])$ is the price for having a short definition of alternative algebras. Zhevlakov conjectured that $J(\text{Alt}[X])$ is a nilpotent ideal. Recently Shestakov and the author managed to prove this conjecture (see [9]); the key moment in the proof is the description of prime alternative superalgebras, given in §1.

THEOREM 2.2. *The radical $J(\text{Alt}[X])$ is nilpotent.*

A similar nilpotence conjecture has also been put forth for the radical of a free Jordan algebra. This radical consists of elements which are identically zero in all special algebras and in the 27-dimensional algebra $H(\mathbf{O}_3)$ (see [18]). It turned out, however, that the situation in Jordan algebras is different from that in alternative algebras. Yu. A. Medvedev has verified that not all elements in the radical are identically zero on the Grassmann envelope of the 10-dimensional exceptional Kac superalgebra F_{10}. It follows from this and from the simplicity of F_{10} that the radical of the free Jordan algebra is not nilpotent.

Another intriguing problem related to infinite-dimensional Jordan algebras is the problem of absolute zero divisors.

An element a of a Jordan algebra J is called an absolute zero divisor if $a^2 = (aJ)a = 0$. In [18] we described the prime Jordan algebras with no nonzero absolute zero divisors. In connection with this it is natural to ask: can a prime Jordan algebra contain nonzero absolute zero divisors? Pchelintsev [19] has constructed a number of fine examples showing that the answer is "yes". We will indicate how one can construct "Pchelintsev monsters" starting from superalgebras.

Let $J = J_0 + J_1$ be an infinite-dimensional Jordan superalgebra of vector fields, $G(J)$ its Grassmann envelope, A a free algebra in the variety generated by $G(J)$. Then A is a prime Jordan algebra with absolute zero divisors. This makes plausible the following

CONJECTURE. Any prime Jordan algebra with a nonzero absolute zero divisor belongs to the variety generated by Grassmann envelope $G(J)$, where J is the Jordan superalgebra of vector fields.

Finally we formulate two theorems on Jordan nil-algebras and Engel Lie algebras which answer some questions of Shirshov [20] and Kostrikin [21].

THEOREM 2.3 [22]. *A Jordan algebra over a field of characteristic zero satisfying* $X^n = 0$ *is solvable.*

THEOREM 2.4 [23]. *A Lie algebra over a field of characteristic zero satisfying* $[\ldots[[x,y],y],\ldots,y] = 0$ *is nilpotent.*

These theorems are consequences of the following two propositions.

PROPOSITION 1. *A Jordan superalgebra whose Grassmann envelope satisfies* $X^n = 0$ *is not prime.*

PROPOSITION 2. *A Lie superalgebra whose Grassmann envelope satisfies* $[\ldots[x,y],\ldots,y] = 0$ *is not prime.*

Bibliography

1. A. I. Kostrikin and I. R. Shafarevich, *Graded Lie algebras of finite characteristics*, Izv. Akad. Nauk SSSR Ser. Mat. **33** (1969), 251–322; English transl. in Math. USSR-Izv. **3** (1969), 237–304.
2. D. A. Leĭtes, *Introduction to the theory of supermanifolds*, Uspekhi Mat. Nauk **35** (1980), 3–57, 255; English transl. in Russian Math. Surveys **35** (1980).
3. V. G. Kac, *Lie superalgebras*, Adv. Math., 1977, v. 26. No. 1. pp. 8–96.
4. M. Scheunert, *The theory of Lie superalgebras*, Lecture Notes in Math. No. 716, Springer-Verlag, Berlin, 1979.
5. A. I. Mal'cev, *Algebraic systems*, "Nauka", Moscow, 1970; English transl., Springer-Verlag, Berlin and New York, 1973.
6. P. Cohn, *Universal algebra*, Harper and Row, New York and London, 1965.
7. C. T. C. Wall, *Graded Brauer groups*, J. Reine Angew. Math. **213** (1964), 187–199.
8. K. A. Zhevlakov, A. M. Slin'ko, I. P. Shestakov, and I. A. Shirshov, *Rings that are nearly associative*, "Nauka", Moscow, 1978; English transl., Academic Press, New York, 1982.
9. E. I. Zel'manov and I. P. Shestakov, *Prime alternative superalgebras and nilpotence of the radical of a free alternative algebra*, Proc. XIVth All-Union Algebraic Conf., L'vov, 1987. (Russian)
10. P. Jordan, J. von Neumann, and E. Wigner, *On an algebraic generalization of the quantum mechanical formalism*, Ann. Math. **36** (1934), 29–64.
11. A. Albert, *On a certain algebra of quantum mechanics*, Ann. Math. **36** (1934), 65–73.
12. V. G. Kac, *Classification of simple Z-graded Lie superalgebras and simple Jordan superalgebras*, Comm. Algebra **13** (1977), pp. 1375–1400.
13. A. R. Kemer, *A remark on the standard identity*, Mat. Zametki **23** (1978), 753–757; English transl. in Math. Notes.
14. H. Weyl, *The classical groups. Their invariants and representations*, Princeton Univ. Press, Princeton, N. J., 1939; 2nd ed., 1946.
15. A. R. Kemer, *Varieties and Z_2-graded algebras*, Izv. Akad. Nauk SSSR **48** (1984), 1042–1059; English transl. in Math. USSR-Izv. **25** (1985), 359–374.
16. A. A. Shestakov, *Radicals and nilpotent elements of free alternative algebras*, Algebra i Logika **14** (1975), 354–365, 370; English transl. in Algebra and Logic **14** (1975).
17. M. Slater, *Prime alternative rings* I, II, J. Algebra **15** (1970), 229–251.
18. E. I. Zel'manov, *Primary Jordan algebras*, Algebra i Logika **18** (1979), 162–175, 253; English transl. in Algebra and Logic **18** (1979).
19. S. V. Pchelintsev, *Prime algebras and absolute zero divisors*, Izv. Akad. Nauk SSSR Ser. Mat. **50** (1986), 79–100; English transl. in Math. USSR-Izv. **28** (1987), 79–98.

20. The Dniester notebook: *Unsolved problems of the theory of rings and modules*, 2nd ed., Inst. Mat. Sibirsk. Otdel. Akad. Nauk SSSR, Novosibirsk, 1976. (Russian)

21. A. N. Kostrikin, *Lie algebras and finite groups*, Trudy Mat. Inst. Steklov **168** (1984), 132–154; English transl. in Proc. Steklov Inst. Math. **3** (1986), 137–161.

22. E. I. Zel'manov, *Solvability of Jordan nil-algebras of bounded index*, Varieties of algebraic systems, Preprint, No. 647, Computing Center Siberian Branch, Akad. Nauk SSSR, 1986. (Russian)

23. ____, *On Engel Lie algebras*, Dokl. Akad. Nauk SSSR **292** (1987), 265–268; English transl. in Soviet Math.-Dokl. **35** (1987), 44–47.

Translated by YU. A. BAKHTURIN

Algebraic Geometry and Combinatorics: a Model-Theoretic Point of View

UDC 510.67 + 514.142 + 512.718

B. I. ZIL'BER

The goal of the present report is to give a survey of some new ideas and constructions originating in the theory of models in the last decade, which are apparently closely connected with interesting phenomena—from our point of view—in algebraic geometry. Moreover these constructions turn out to be quite useful in classical combinatorics and have led to noteworthy new results in that context. Therefore, another goal of this survey is to demonstrate the connection between algebraic geometry and combinatorics which has been revealed by a model theoretic approach. The theme of the present report is close to the theme of the survey [18], but here we confine ourselves to a more algebraic presentation.

The well-known concept of a pregeometry (or matroid), stemming from the thirties [1], provides a basis for a general point of view in the three aforementioned areas of mathematics. A set M with an operation cl: $2^M \to 2^M$ is called a *pregeometry*, if the following conditions are satisfied for all $X, Y \subseteq M$ and $x, y, z \in M$:

(1) $X \subseteq \mathrm{cl}(Y) \Rightarrow \mathrm{cl}(X) \subset \mathrm{cl}(Y)$; $X \subseteq \mathrm{cl}(X)$;
(2) $\mathrm{cl}(X) = \bigcup\{\mathrm{cl}(X'): X' \subseteq X, X' \text{ finite}\}$;
(3) $z \in \mathrm{cl}(X, y) - \mathrm{cl}(X) \Rightarrow y \in \mathrm{cl}(X, z)$.

This definition provides a decent generalization of the notion of algebraic closure in fields and linear closure in vector spaces. A new ingredient is the condition of *homogeneity* for a pregeometry:

(4) If $y_1, y_2 \notin \mathrm{cl}(X)$, then there is an automorphism α of the pregeometry (M, cl) such that $\alpha|_X = \mathrm{id}|_X$ and $\alpha y_1 = y_2$.

We also recall that a pregeometry is called a *geometry* if:

(5) $\mathrm{cl}(\{x\}) = \{x\}$ for all $x \in M$, and $\mathrm{cl}(\varnothing) = \varnothing$.

1980 *Mathematics Subject Classification* (1985 *Revision*). Primary 05B35; Secondary 03C45, 03C60, 05B25, 14A25.

We will denote by $\dim X$ the maximal number of elements of X which are independent in the sense of cl.

It is easy to see that a vector space $V^n(K)$, a projective space $P^n(K)$, and an affine space $A^n(K)$ over any skew field K are all homogeneous pregeometries, where we take as $\operatorname{cl}(X)$ the liner span of the set X. Another important example is an algebraically closed field F with the algebraic closure operator. Still another example is a degenerate geometry: a set S with the trivial closure operator, $\operatorname{cl}(X) = X$ for all X. Among the examples cited, $P^n(K)$, $A^n(K)$, and S are all geometries.

We remark that for $\dim M \geq 6$ there are no other known examples of homogeneous geometries M, essentially different from those cited. More precisely, for all known examples, by going from the pregeometry M to the quotient geometry $\widehat{M}$ (by the natural factorization [1]) we obtain S, $P^n(K)$, $A^n(K)$, or $\widehat{F}$ (here n may be an infinite cardinal). Without the homogeneity condition there are known examples of other types.

Interest in homogeneous geometries in the theory of models arises in connection with the important role which minimal structures play in the theory of categoricity and stability [18].

We recall that a *structure* (or algebraic system) $\mathfrak{M}$ is a set M together with relations $R_1, \ldots, R_k, \ldots$ and operations $f_1, \ldots, f_m, \ldots$ given on M:

$$\mathfrak{M} = \langle M; R_1, \ldots, R_k, \ldots, f_1, \ldots, f_m, \ldots \rangle.$$

The structure $\mathfrak{M}$ is called *minimal* if M is infinite and any definable subset of the set M (in the first order language with parameters from M) is either finite or has finite complement. We specify a closure operator cl on M: for $X \subseteq M$, we consider the group $\operatorname{Aut}_X(\mathfrak{M})$ of automorphisms of $\mathfrak{M}$ which fix X pointwise, and we set $\operatorname{cl}(X) = $ the union of all finite orbits of $\operatorname{Aut}_X(\mathfrak{M})$ on M. For such an operation cl given in this way on a minimal structure $\mathfrak{M}$, (M, cl) will be a homogeneous pregeometry.

The canonical example of a minimal structure is an algebraically closed field $(F; +, \cdot)$. Indeed, an arbitrary definable subset of F is the set of roots of some polynomial, or the complement of such a set, by Tarski's theorem. The infinite structures $V^n(K)$, $P^n(K)$, and $A^n(K)$—with a natural choice of relations and operations—are also minimal.

The methods and ideas of the theory of models, applied in the investigation of minimal structures and described below, led to the following theorem.

THEOREM 1. *A homogeneous locally finite geometry of dimension at least 7 whose lines contain at least three points is isomorphic with $P^n(K)$ or $A^n(K)$ for some finite field K.*

Here the local finiteness means that $\operatorname{cl}(X)$ is finite for any finite set X. In particular, a locally finite geometry of finite dimension is finite.

This theorem was considered as a conjecture in [3] from the point of view of model theory, and independently, in another but equivalent formulation from the point of view of finite combinatorics and group actions (cf. [15]). Its proof was obtained by the author of this survey in 1980 for the case of infinite dimension [4], [5], [6], and at the same time G. Cherlin and others remarked that the same version of the theorem can be obtained from a complete classification of all doubly transitive finite groups, following from the classification of the simple groups [10]. Using the ideas of [3] and [4], D. Evans carried out a proof of Theorem 1 in the infinite-dimensional case by combinatorial methods [11], and later a finite variant for dimension at least 23 (still unpublished). In 1986 the author obtained the present version (to appear) by modifying the methods of [3]–[6].

Below we deal with the development of the methods of [3]–[6].

Since the theorem stated above, and our commentary, might give the impression that it is possible to replace these methods by finite combinatorial computations, we will mention a result not reducible to finite combinatorics [7].

Let G be a group acting on a set M (possibly infinite). For any $X \subseteq M$ we write:

$$G_X = \{g \in G : \forall x \in X, gx = x\}$$
$$\mathrm{cl}_G(X) = \{y \in M : \forall g \in G_X, gy = y\}.$$

We say that the group G is *hereditarily transitive* on M if for all $X \subseteq M$ the subgroup G_X acts transitively on $M - \mathrm{cl}_G(X)$.

It is easy to see that if G is hereditarily transitive on M, then (M, cl_G) is a homogeneous pregeometry.

THEOREM 2. *If G is a hereditarily transitive group on M, $\dim M = n \geq 8$, and G is not 3-transitive on $M - \mathrm{cl}_G(\varnothing)$, then there is a vector space V over a skew field and a subspace W such that G can be identified with a "large" subgroup of the group $\mathrm{GL}_W(V)$, or the group $\mathrm{AGL}_W(V)$, where $\mathrm{GL}_W(V) = \{g \in \mathrm{GL}(V) : g|_W = \mathrm{id}|_W\}$, and $\mathrm{AGL}_W(V) = T(V) \rtimes \mathrm{GL}_W(V)$, with $T(V)$ the group of parallel translations of the space V. Under this identification the action of G on $M - \mathrm{cl}(\varnothing)$ agrees with the action of $\mathrm{GL}_W(V)$ on $V - W$ or of $\mathrm{AGL}_W(V)$ on V.*

To say that G is a "large" subgroup of $\mathrm{GL}_W(V)$ means that for any two systems $(v_1, \ldots, v_n)$ and $(v_1', \ldots, v_n')$ of linearly independent vectors modulo W, there is a unique $g \in G$ such that $gv_1 = v_1', \ldots, gv_n = v_n'$, and similarly in the case of $\mathrm{AGL}_W(V)$. For finite G it follows from the theorem of Cameron and Kantor [9] that $G = \mathrm{GL}_W(V)$ or $G = \mathrm{AGL}_W(V)$. In the case that V is a vector space over an infinite skew field, and n is finite, we do not know if a "large" subgroup can be proper. We remark that the answer is positive when G is the group of automorphisms of a so-called v^*-algebra [13], [14], and this essentially gives a new proof of a well-known theorem of Urbanik describing v^*-algebras [17].

We turn to the consideration of the methods applied to obtain a number of results on homogeneous pregeometries, in particular Theorems 1 and 2.

The basis of these methods is the familiar notion of (formular) definability. Let $\mathfrak{M} = \langle M; R_1, \ldots, R_k, \ldots, f_1, \ldots, f_m, \ldots \rangle$ be a structure. We say that a subset $S \subseteq M^k$ is *definable* in the structure $\mathfrak{M}$ if there is a formula $\varphi(x_1, \ldots, x_k)$ in the corresponding first order language constructed from $R_1, \ldots, f_1, \ldots$ and names for the elements of M (parameters), such that

$$\langle s_1, \ldots, s_k \rangle \in S \Leftrightarrow \mathfrak{M} \models \varphi(s_1, \ldots, s_k).$$

We say that the structure $\mathfrak{N} = \langle N; Q_1, \ldots, g_1, \ldots \rangle$ is *definable* in the structure $\mathfrak{M}$ if the universe of N, the predicates $Q_1, \ldots$, and the graphs of the operations $g_1, \ldots$ are all quotient sets of the form S/E with S a definable subset of some power M^k and with E a definable equivalence relation on S ($E \subseteq M^{2k}$).

In some important cases, for example when M is finite, the notion of definability of $S \subseteq M^k$ using the parameters $m_1, \ldots, m_n \in M$ is equivalent to the condition of invariance under those automorphisms of the structure $\mathfrak{M}$ which fix the parameters $m_1, \ldots, m_n$. In all cases the latter condition is a consequence of definability.

EXAMPLE. Any group $G(F)$ of the F-rational points of an algebraic group G is definable in the field structure $\mathfrak{F} = \langle \mathfrak{F}; +, \cdot \rangle$.

We will show this in the example of the group $\mathrm{PSL}(2, F) = \langle G, \cdot \rangle$. The universe of G is obtained by taking the quotient of the set

$$S = \{\langle s_{11}, s_{12}, s_{21}, s_{22} \rangle : s_{11} \cdot s_{22} - s_{12} \cdot s_{22} = 1\}$$

modulo E:

$$\langle s_{11}, s_{12}, s_{21}, s_{22} \rangle E \langle s'_{11}, s'_{12}, s'_{21}, s'_{22} \rangle \Leftrightarrow (\exists x)(s'_{11} = x s_{11} \& \cdots \& s'_{22} = x s_{22}).$$

Similarly one may give a suitable formula for the graph of the multiplication operation as a subset of S^3 (that is as a subset of F^{12}), and the graph of "$\cdot$" is obtained by factoring by E^3.

A special type of structure plays an important role below.

A *pseudoplane* is defined as a structure $\mathfrak{P}$ with two one-place predicates P and L and one binary relation I, where the universe is partitioned by the predicates P and L into two disjoint subsets—the set P of "points" and the set L of "lines"—and $I \subseteq P \times L$, with xIy interpreted as "the point x lies on the line y". Here it is assumed that the following conditions are satisfied:

Through any two points there pass only finitely many lines;
Two lines have only a finite intersection;
Infinitely many lines pass through any given point;
Infinitely many points lie on any given line.

It is easy to see that an infinite affine or projective plane is a pseudoplane.

However, more typical examples may be obtained as follows:

Let F be an algebraically closed field, P the set of F-points of an algebraic surface of dimension 2 over F, and L a two-dimensional algebraic family of irreducible curves on P. Let P' be the set of those points of P through which infinitely many curves from L pass, and L' the set of those lines of L whose intersection with P' is infinite. Then (P', L') is a pseudoplane with respect to the natural relation "a point lies on a curve". Moreover the given pseudoplane is definable in the structure $\langle F; +, \cdot \rangle$.

We call a structure $\mathfrak{M}$ *rather rich* if a pseudoplane is definable in $\mathfrak{M}$.

Now we are able to reformulate the idea of our approach to the study of homogeneous pregeometries.

THEOREM 3 [19]. *Let $\mathfrak{M}$ be a minimal structure and $\dim M$ infinite with respect to the pregeometry (M, cl). Then $\mathfrak{M}$ is rather rich if and only if its pregeometry is not reducible to a degenerate, affine, or projective geometry.*

Modulo Theorem 3 the proof of theorems of the type of Theorems 1 and 2 reduces to the proof of the nonexistence of pseudoplanes under specified conditions. As such a condition we may take local finiteness as in Theorem 1, or a special type of cl operation, as in Theorem 2.

In other words, the use of pseudoplanes definable in a homogeneous pregeometry is the key to the problem. The connection between pseudoplanes and homogeneous pregeometries gives a very important tool: the notion of rank. This notion is the strict analog of the notion of dimension of an algebraic variety. We give the rank only for definable subsets of a special form, so as not to pile up technical details.

Accordingly, let (M, cl) be a homogeneous pregeometry, $A \subseteq M^k$, A a set which is definable in M without parameters. We set:

$$\mathrm{rank}\, A = \max\{\dim\{a_1, \ldots, a_k\}: \langle a_1, \ldots, a_k \rangle \in A\}.$$

Clearly, this definition is compatible with the definition of dimension of an affine variety A over a field, if we bear in mind that $\dim\{a_1, \ldots, a_k\}$ is the transcendence degree of the field extension generated by $a_1, \ldots, a_k$.

Thus the rank so defined has a number of essential properties in the case of an infinite dimension M:

(1) $\mathrm{rank}\, A = 0$ iff A is finite;
(2) $\mathrm{rank}\, A \geq n + 1$ iff there is a sequence of definable subsets $A \supset A_1 \supset A_2 \supset \cdots \supset A_k \supset \cdots$ so that $\mathrm{rank}(A_{i+1} - A_i) \geq n$;
(3) if $f: A \to B$ is a definable map, then

$$\mathrm{rank}\, A \leq \mathrm{rank}\, B + \max\{\mathrm{rank}\, f^{-1}(b): b \in B\}.$$

We remark that the rank is easy to use in infinite dimensional pregeometries, but with care it may be adapted for effective use even in finite pregeometries. Thus we gain the ability to speak even about finite pseudoplanes;

in place of a condition of the type "A is infinite," in the definition of a pseudoplane we may say "rank $A > 0$". In fact Theorem 1 follows from Theorem 3 and

THEOREM 1'. *There is no rather rich finite pregeometry.*

Borovik [2] proposes to call structures in which a rank function on the definable sets is given, satisfying axioms (1)–(3), *structures with dimension*. We note that an abstract infinite structure $\mathfrak{A}$ with dimension is always connected with at least one homogeneous pregeometry, more precisely with some minimal structure. For this, in order to construct this minimal structure, it is sufficient to find a definable set M in $\mathfrak{A}$ such that rank $M = 1$, and M is not divisible into two subsets of rank 1. Considering the set M together with the relations and definitions induced by the structure $\mathfrak{N}$, we obtain a minimal structure $\mathfrak{M}$.

We call a structure with dimension $\mathfrak{N}$ *strongly categorical*, if any automorphism of $\mathfrak{N}$ inducing the identity on some minimal structure $\mathfrak{M}$ defined in $\mathfrak{N}$ is given by a definable map.

EXAMPLE. A simple group with dimension is rather rich and strongly categorical. All algebraic groups over algebraically closed fields are groups with dimension.

THE CENTRAL CONJECTURE. (A) In any rather rich structure $\mathfrak{N}$ with dimension, some infinite field $\mathfrak{F} = \langle F; +, \cdot \rangle$ is definable.

(B) Conversely, if $\mathfrak{N}$ is strongly categorical, then $\mathfrak{N}$ is definable in $\mathfrak{F}$.

We remark that the field $\mathfrak{F}$ in this conjecture may be assumed to be algebraically closed as well, in view of the well-known

THEOREM 4 [12]. *Any field with dimension is algebraically closed.*

We give great weight to the Central Conjecture, since a positive answer would give the classification of minimal structures of large dimension (and possibly even all homogeneous geometries), and also would characterize structures with dimension to a remarkable degree. In particular, it would follow from a positive answer, in conjunction with some known facts, that simple groups with dimension are exactly the groups of F-rational points of simple algebraic groups (a well-known conjecture of Cherlin and the author).

On the other hand, the Central Conjecture expresses the idea that algebraic geometry over algebraically closed fields is reducible to the study of a rather rich structure with dimension. That is, if the algebraic structure on such a structure is rather rich, then it is possible to recover the initial field, and then even all objects given by algebraic geometry, under the assumption of strong categoricity. Without this assumption the situation is somewhat complicated, but sufficiently under control.

Unfortunately, there are very few positive results in the direction of the Central Conjecture. Attention should be given to the special case of this conjecture in which $\mathfrak{N}$ is automatically definable in $\mathfrak{F}$, that is when $\mathfrak{N}$ is essentially an object arising in algebraic geometry. In this case part (B) of the conjecture falls away, and the problem reduces to the definability of $\mathfrak{F}$ in $\mathfrak{N}$. We illustrate the possible applications.

EXAMPLE [16], [20]. Let G be the group of F-rational points of a simple algebraic group over an algebraically closed field F. Then there is a field structure $\mathfrak{F}$ definable in G.

(The proof does not require any structural results concerning G beyond the nonnilpotency of Borel subgroups.)

As remarked above, G is strongly categorical. Therefore it follows that any automorphism of G is the composition of automorphisms induced by an automorphism of $\mathfrak{F}$, and definable automorphisms of G. With the help of Tarski's theorem (the constructibility of sets definable in $\mathfrak{F}$) it may be shown that a definable automorphism of G is the composition of a rational automorphism and Frobenius automorphisms. As a result a version of the well-known theorem of Borel-Tits [8] is obtained for groups over algebraically closed fields.

We consider now still another special type of structures with dimension, which are objects occurring in algebraic geometry. Let M be the set of F-rational points of a smooth irreducible algebraic curve of degree d over an algebraically closed field F, and $R(x_1, x_2, x_3)$ the following ternary relation on M: "the points x_1, x_2, x_3 are collinear". Then the structure $\mathfrak{M} = \langle M; R \rangle$ is a structure with dimension, and furthermore is a minimal structure. In the context of the Central Conjecture great importance is attached to the question:

For which curves is the structure $\mathfrak{M} = \langle M; R \rangle$ *rather rich?*

If the degree d of the curve is 2, the relation R is degenerate. In the case $d = 3$, as is well known, it is possible to give M the structure of an abelian group $(M; +)$ (generally speaking, the operation $+$ is given for "almost all" pairs of summands) so that $(M; R)$ is definable in $(M; +)$, and conversely. Thus it follows that the pregeometry of the structure $(M; R)$ is reducible to projective geometry over some subfield F_0 of the field F. In other words, for $d \leq 3$ $(M; R)$ is not rather rich.

It seems very likely that for $d \geq 4$ $(M; R)$ is rather rich, and together with the Central Conjecture this would imply that there is a field structure $\mathfrak{F}$ definable in $(M; R)$.

We now postulate the correctness of this assertion. Then the considerations applied in the discussion of the foregoing example lead us to state the following conjecture:

If $\varphi: (M_1; R) \to (M_2; R)$ *is an isomorphism of structures, where* M_1 *and* M_2 *are the* F-*rational points of a smooth algebraic curve of degree* $d \geq 4$, *then* φ *can be represented as the composition of an isomorphism induced by an isomorphism of the field* $\mathfrak{F}$ *and a biregular isomorphism of curves.*

In conclusion we remark that the latter theme can be extended significantly. It is appropriate to take for M any variety. Of course, in place of R any algebraic relation on M may be considered, and also several relations simultaneously.

Bibliography

1. Martin Aigner, *Combinatorial theory*, Springer-Verlag, 1979.
2. A. V. Borovik, *Theory of finite groups, and uncountably categorical groups*, Preprint No. 511, Computing Center, Siberian Branch of the Academy of Sciences of the USSR, Novosibirsk, 1984. (Russian)
3. B. I. Zil'ber, *Strongly minimal countably categorical theories*, Sibirsk. Mat. Zh. **21** (1980), no. 2, 98-112; English transl. in Siberian Math. J. **21** (1980), no. 2, 219-230.
4. ____, *Totally categorical structures and combinatorial geometries*, Dokl. Akad. Nauk SSSR **259** (1981), no. 5, 1039-1041; English transl. in Soviet Math. Dokl. **24** (1981), no. 1, 149-151 (1982).
5. ____, *Strongly minimal countably categorical theories.* II, Sibirsk. Mat. Zh. **25** (1984), no. 3, 71-88; English transl. in Siberian Math. J. **25** (1984), no. 3, 396-412.
6. ____, *Strongly minimal countably categorical theories.* III, Sibirsk. Mat. Zh. **25** (1984), no. 4, 63-77; English transl. in Siberian Math. J. **25** (1984), no. 4, 559-571.
7. ____, *Hereditarily transitive groups and quasi-Urbanik structures*, Trudy Inst. Mat. (Novosibirsk) **8** (1988), Teor. Model. i ee Primenen., 58-77. (Russian)
8. A. Borel and J. Tits, *Homomorphisms "abstraits" de groupes algébriques simples*, Ann. of Math. **97** (1973), 499-571.
9. P. J. Cameron and W. M. Kantor, *2-transitive and antiflag transitive collineation groups of finite projective spaces*, J. Algebra **60** (1979), 384-422.
10. G. Cherlin, L. Harrington, and A. H. Lachlan, $\aleph_0$-*categorical*, $\aleph_0$-*stable structures*, Ann. Pure and Appl. Logic **28** (1985), no. 2, 103-136.
11. D. M. Evans, *Homogeneous geometries*, Proc. London Math. Soc. (3) **52** (1986), no. 2, 305-327.
12. A. Macintyre, *On* $\aleph_1$-*categorical theories of fields*, Fund. Math. **71** (1971), 1-25.
13. E. Marczewski, *Independence in some classes of abstract algebras*, Bull. Acad. Polon. Sci. Ser. Math. **7** (1959), 611-616.
14. W. Narkiewicz, *Independence in a certain class of abstract algebras*, Fund. Math. **50** (1961), 333-340.
15. P. Neumann, *Some primitive permutation groups*, Proc. London Math. Soc. **50** (1985), 265-281.
16. B. Poizat, *MM. Borel, Tits, Zilber et le nonsense général*, J. Symbol. Logic **53** (1988), 124-131.
17. K. Urbanik, *A representation theorem for* v^*-*algebras*, Fund. Math. **52** (1963), 291-317.
18. B. I. Zil'ber, *The structure of models of uncountable categorical theories*, In: Proc. Int. Congr. Math. Warszawa, 1982, Warszawa, Amsterdam, 1984, Vol. 1, pp. 359-368.
19. ____, *Structural properties of models of* $\aleph_1$-*categorical theories*, In: Proc. 7th Int. Congr. IMPS. Salzburg, 1983, Amsterdam, 1985, pp. 115-128.
20. ____, *Some model theory of simple algebraic groups over algebraically closed fields*, Colloq. Math. **48** (1984), No. 2, 173-180.

Translated by G. L. CHERLIN

Translator's note. The Central Conjecture was subsequently refuted by a construction of Hrushovski, and verified for a large class of structures interpretable in an algebraically closed field by Rabinovich. The conjecture regarding simple groups with dimension remains unresolved.

Finiteness Theorems for Limit Cycles

YU. S. IL'YASHENKO

1. Statement of results

THEOREM 1. *A polynomial vector field on the real plane has only a finite number of limit cycles.*

THEOREM 2. *An analytic vector field on a closed 2-manifold has only a finite number of limit cycles.*

THEOREM 3. *A singularity of an analytic vector field on the real plane always has a neighborhood disjoint from limit cycles.*

It has been known since the time of Poincaré and of Dulac (see [4] or [5], for example) that this theorem is a consequence of the following one:

THEOREM 4. *An elementary separatrix cycle* * *of an analytic vector field on a 2-manifold always has a neighborhood disjoint from limit cycles.*

Recall that a *separatrix cycle* is a polygon composed of separatrices. (Precise definitions of all terms presented in this paper at an intuitive level can be found in [1] or [5], for example.) A separatrix cycle is *elementary* if all the singularities belonging to it are elementary, that is they have at least one nonzero eigenvalue.

The monodromy map of a separatrix cycle is defined just as for an arbitrary cycle, the only difference being that instead of a transversal we take a semi-transversal, that is a half-interval transverse to the cycle (Figure 1). It is convenient to regard monodromy maps as map germs $(\mathbb{R}^+, 0) \to (\mathbb{R}^+, 0)$.

THEOREM 5 (Identity Theorem). *If the monodromy map of a separatrix cycle of an analytic vector field has countably many fixed points then it is the identity map.*

1980 *Mathematics Subject Classification* (1985 *Revision*). Primary 34C25; Secondary 58F21.
* *Editor's note.* The Russian is сложный цикл (literally *compound cycle*).

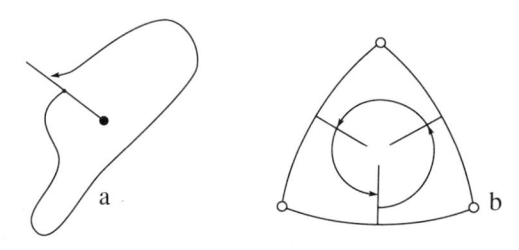

FIGURE 1

Theorem 4 clearly follows from Theorem 5. The remainder of the paper is devoted to the basic ideas of the proof of the Identity Theorem, namely superexact asymptotic series (§§4 and 5) and the continuation of monodromy maps to the complex domain (§§6–8).

Consider the class of those maps that arise as monodromy maps of separatrix cycles. It is not possible to characterize this class either in terms of properties of the maps that belong to it, or from their definition. This is connected primarily with the presence of saddles whose ratio of eigenvalues has "poor arithmetic"; for more details see [5], Chapter 1, §5.

2. Sketch of the proof

To prove the Identity Theorem we construct a class of map germs $(\mathbb{R}^+, 0) \to (\mathbb{R}^+, 0)$ that contains the monodromy maps, and such that the germs belonging to this class have two properties: *decomposability* and *continuability*.

Decomposability means that to each germ there corresponds a series containing information on not only the polynomial but also the exponential asymptotics of the germ. These series are called STAR [acronym from Russian words for superexact asymptotic series]; see §§4 and 5 for more details. The triviality of such a series, that is its collapse to a single term x, indicates that the error vanishes very rapidly as $x \to 0$. Moreover, the terms of the series do not oscillate, so that the presence of a countable number of fixed points for the germ implies the triviality of the corresponding series.

Continuability means that the germ can be continued in the complex domain to a "map cochain"—a piecewise continuous map holomorphic away from cut lines. For those map-cochains arising from the continuation of a monodromy map a theorem of Phragmén-Lindelöf type holds: if the error for the germ vanishes too rapidly then it is identically zero. Triviality of the STAR immediately guarantees this "too rapid" vanishing. We obtain the following implication:

$$\Delta_\gamma \in \text{Fix}_\infty \Rightarrow \hat{\Delta}_\gamma = x \Rightarrow \Delta_\gamma - x \equiv 0$$

where Δ_γ is the monodromy map for the separatrix cycle γ, Fix_∞ is the set of germs with countably many fixed points, and $\hat{\Delta}_\gamma$ is the STAR for Δ_γ.

For elementary separatrix cycles with nondegenerate singularities (they will all be hyperbolic saddles if the cycle is not a single point, so this case is called hyperbolic) this program was carried out in [4] using usual series, not superexact ones. In the general case, in order to describe analytic continuation of the monodromy map we use the geometric theory of normal forms for resonant vector fields and maps as in [1], [2], [3], [14], or [15].

In the next section we describe the polynomial asymptotics of monodromy maps and explain why they are insufficient to describe the maps completely. §§4 and 5 are devoted to STAR. In §§6–8 we describe the continuation to the complex domain of the germs whose composition gives the monodromy maps. It is the proof of the Identity Theorem for the composition of these germs that presents the basic difficulty; this part of the proof will not be presented here. Note that in the memoir of Dulac, on the other hand, it is the study of correspondence maps that is fundamental and their compositions are expanded in asymptotic series in an elementary fashion.

3. Dulac series and exponentially small errors for monodromy maps

Dulac [10] studied asymptotic series for monodromy maps of separatrix cycles of analytic vector fields, the series approximating the maps to an arbitrarily high degree.

DULAC'S THEOREM. *Given a separatrix cycle of an analytic vector field, it is possible to choose a semi-transversal and a coordinate chart on it such that the monodromy map is either a flat function, or the inverse of a flat function, or is semiregular. Semiregularity means that the germ can be expanded as a Dulac asymptotic series*:

$$cx^{\nu_0} + \sum_1^\infty P_j(\ln x) x^{\nu_j}, \qquad \nu_j > 0, \qquad \nu_j \nearrow \infty,$$

where $c > 0$ and the P_j are polynomials (*the arrow $\nearrow$ denotes a monotonically increasing sequence tending to a limit*). *The partial sums of the Dulac series approximate the monodromy map up to arbitrary degree in* x.

This theorem is valid not only for analytic vector fields, but for smooth ones [5] as well. Therefore a theorem on finiteness of limit cycles cannot be deduced from it alone. What is at least clear is that if closed orbits accumulate on a separatrix cycle then the monodromy map has a countable number of fixed points accumulating on zero, and its Dulac series is equal to x (see [10], §23, and §3 of the Introduction to [5]).

In the hyperbolic case the monodromy map is determined uniquely by its Dulac series. This means that if the monodromy maps of separatrix cycle of (different) analytic vector fields with nondegenerate singularities have identical Dulac series then they coincide. This was essentially proved in [4]. In particular, the following theorem holds.

THEOREM. *If the Dulac series for the monodromy map of a separatrix cycle of an analytic vector field with nondegenerate singularities is equal to x, then the monodromy map is the identity.*

This implies the Identity Theorem (and hence also Theorems 1–4) for vector fields with nondegenerate singularities (Theorem 3 in this case is trivial).

In the general case, however, the Dulac series does not determine the monodromy map. There exists a separatrix cycle of an analytic vector field containing two degenerate elementary singularities whose monodromy map differs from the identity by a nonzero flat term of order $\exp(-1/x)$ (see [4]).

Usual asymptotic series are likewise insufficient to determine the monodromy map uniquely.

4. Superexact asymptotic series (STAR)

Consider a set M_1 of map germs $(\mathbb{R}^+, 0) \to (\mathbb{R}^+, 0)$. Each germ can be expanded as an asymptotic series (Dulac series, for example) whose partial sums approximate the germ up to arbitrary degree in x. We call these the "usual" series. However, it is desirable to expand the germs as series whose terms have not only polynomial but also exponential order of smallness. At first sight this is impossible: whatever partial sum of a usual asymptotic series we take, the difference between this sum and the approximated function has in general a polynomial order of smallness and to take account of exponentially vanishing terms seems meaningless.

This difficulty can be circumvented as follows. We introduce an intermediate class M_0 of functions each of which can be expanded as a usual asymptotic series. Then the germs belonging to M_1 are expanded as series in exponentially vanishing functions, where the coefficients are now not numbers but functions of the intermediate class. The simplest example of a STAR looks as follows:

$$\Sigma = a_0(x) + \sum a_j(x)\exp(-\nu_j/x), \qquad a_j \in M_0, \qquad \nu_j \nearrow \infty. \qquad (*)$$

The series Σ is said to be asymptotic for the germ f if for any $\nu > 0$ there is a partial sum Σ_n of Σ such that

$$f - \Sigma_n = O(\exp(\nu/x)).$$

All the information on the expansion of f as a usual asymptotic series is contained in the free term of the superexact one (the usual series for a_0 and f coincide).

REMARK. Superexact asymptotic series apparently arise for many classes of maps, and monodromy maps are not the simplest of them.

5. Superexact series and fixed points

We show by a simple example how to use superexact series to prove the Identity Theorem.

LEMMA. *Let M_1 and M_0 be two classes of map germs $(\mathbb{R}^+, 0) \to (\mathbb{R}^+, 0)$ and function germs $(\mathbb{R}^+, 0) \to (\mathbb{R}^+, 0)$, respectively. Suppose the germs in these classes are bounded below in the following sense:*

any nonzero germ of M_0 vanishes more slowly than some power of x;

the error for a nonzero germ of M_1 vanishes more slowly than some exponential $\exp(-\nu/x)$.

More precisely,

$$f \in M_0, \ |f| < x^\nu \ \forall \nu > 0 \Rightarrow f \equiv 0;$$
$$f \in M_1, \ |f - x| < \exp(-\nu/x) \ \forall \nu > 0 \Rightarrow f - x \equiv 0.$$

Suppose in addition that every germ belonging to M_0 can be expanded as a Dulc series (in which the leading term cx^{ν_0} may be zero, in contrast to §3), and that every germ belonging to M_1 can be expanded as a STAR (∗) with coefficients $a_j \in M_0$, $j = 0, 1, \ldots$. Then the Identity Theorem holds for germs of the class M_1, that is

$$f \in M_1 \cap \mathrm{Fix}_\infty \Rightarrow f = \mathrm{id}.$$

The lemma is proved by the same procedure as Dulac's lemma [10], [5], which may be stated as follows: if a function germ can be expanded as a Dulac asymptotic series and has a countable number of zeros then its Dulac series is identically zero.

Assume the lemma is false: $f \in M_1 \cap \mathrm{Fix}_\infty$, $f \neq \mathrm{id}$. Let (∗) be a STAR for f.

Suppose first that $a_0 \neq \mathrm{id}$ in the formula (∗). Then the Dulac series for $a_0 - \mathrm{id}$ is nonzero, otherwise the germ $a_0 - \mathrm{id}$ vanishes faster than any power and so by the hypothesis of the lemma it is identically zero. Hence a_0 is equal to the leading term of its Dulac series multiplied by $(1 + o(1))$. Therefore there exists ν such that

$$|a_0 - \mathrm{id}| > x^\nu.$$

Moreover, the STAR expansion (∗) of f implies:

$$|f - \mathrm{id}| \geq |a_0 - \mathrm{id}| + -(a_1 \exp(-\nu_1/x))(1 + o(1)) > x^\nu(1 + o(1)).$$

Hence $f - \mathrm{id} \neq 0$ for small $x = 0$; $f \in \mathrm{Fix}_\infty$. Contradiction.

Suppose now that $a_0 = \mathrm{id}$, $f \neq \mathrm{id}$. Then the STAR (∗) for f is distinct from id, otherwise we would have $f - \mathrm{id} = O(\exp(-\nu/x))$ for any $\nu > 0$ and so $f = \mathrm{id}$ by the hypothesis of the Lemma. From the formula (∗) and the definition of that series expansion we have

$$f - \mathrm{id} = (a_1 \exp(-\nu_1/x))(1 + o(1)),$$

$a_1 \in M_0$, $a_1 \neq 0$. Arguing as in the previous paragraph we obtain that $a_1 \neq 0$ for small $x \neq 0$. The two other factors in the expansion for $f_0 - \mathrm{id}$ also do not vanish close to zero. Therefore f has no fixed points close to zero apart from zero itself: contradiction.

This argument serves as a model for the proof of the Identity Theorem. Monodromy maps for separatrix cycles can be expanded as superexact asymptotic series not only in terms of simple exponentials but also multiple ones of the form

$$\exp(-\exp \circ \cdots \circ \exp(1/x)).$$

When the coefficients of these series are continued into the complex domain we obtain so-called *functional cochains*: piecewise continuous functions that are holomorphic on their set of continuity. They also arise naturally when correspondence maps for hyperbolic sectors of elementary singularities (see Figure 2) are continued into the complex domain. We describe these continuations for the hyperbolic and then the nonhyperbolic case.

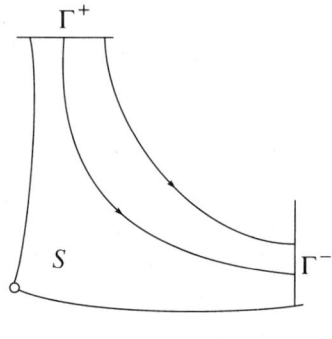

FIGURE 2

6. Correspondence maps in the hyperbolic case

These maps are continued into the complex domain as so-called "almost regular germs" defined as follows. Consider the class of "standard regions"

$$\Omega_c = \{\xi \geq c(1 + \eta^2)^{1/4}\}$$

in the plane of the complex variable $\zeta = \xi + i\eta$. Let $\mathbb{C}^+$ denote the half-plane $\xi \geq 0$. Two holomorphic maps of standard regions are said to be equivalent if one of them is a continuation of the other. A germ $(\mathbb{C}^+, \infty) \to (\mathbb{C}^+, \infty)$ is by definition an equivalence class of maps biholomorphic onto their image and defined on standard regions. Maps in this class are called representatives of the germ. A map germ $(\mathbb{C}^+, \infty) \to (\mathbb{C}^+, \infty)$ is said to be almost regular if it can be expanded as an "exponentially asymptotic Dulac series":

$$\Sigma = \nu\zeta + c + \sum Q_j(\zeta)\exp(-\nu_j\zeta),$$

where $\nu > 0$, $c \in \mathbb{R}$, the Q_j are real polynomials, and $\nu_j > 0$. The expansion is understood in the following sense: there exist a standard region Ω and a representative f of the germ such that for every $\nu > 0$ there exists a partial sum of the series Σ approximating f in Ω with accuracy up to $O(\exp(-\nu\xi))$ as $\zeta \to \infty$ in Ω.

We say that a map germ $g\colon (\mathbb{R}^+, 0) \to (\mathbb{R}^+, 0)$ can be continued in a logarithmic chart to an almost regular germ if
$$f\colon (\mathbb{R}^+, \infty) \to (\mathbb{R}^+, \infty), \qquad f = h^{-1} \circ g \circ h,$$
(where $h\colon (\mathbb{R}^+, 0) \to (\mathbb{R}^+, \infty)$, $x \mapsto \xi = -\ln x$) can be continued to an almost regular germ $(\mathbb{C}^+, \infty) \to (\mathbb{C}^+, \infty)$.

LEMMA. *Let v be the germ of an analytic vector field on the plane with a singularity of "hyperbolic saddle" type. Suppose coordinate charts are taken on incoming and outgoing semitransversals of the hyperbolic saddle that are non-negative and can be continued to analytic charts on the transversals, where the coordinate 0 corresponds to the end points of the semitransversals. Then the correspondence map becomes the germ of a map $(\mathbb{R}^+, 0) \to (\mathbb{R}^+, 0)$. We assert that this germ is almost regular.*

This lemma is a simple modification of Basic Lemma 3 of [4].

We next describe correspondence maps for degenerate elementary singularities, following [6] and [13]. First we define normalizing map-cochains.

7. Normalizing map-cochains

DEFINITION. Let $f\colon (\mathbb{C}, 0) \to (\mathbb{C}, 0)$ be the germ of a conformal map not equal to the identity but with linear part the identity, and let $S \subset \mathbb{C}$ be a sector with vertex 0. A map germ $h\colon (S, 0) \to (\mathbb{C}, 0)$, holomorphic within S and continuous at zero, is called *normalizing* for f if h is a conjugacy between f and the unit time map along the orbits of a vector field holomorphic at zero. Representaties of a normalizing germ are called normalizing maps.

A dissection of a punctured disc with center at 0 in the plane, with $2k$ equal sectors one of whose bounding radii lies on the positive real axis, is called a *good k-dissection*.

THEOREM (on sectorial normalization [2], [11], [14]). *Let*
$$f\colon z \mapsto z + az^{k+1} + \cdots, \qquad a = i\omega \neq 0, \qquad \omega \in \mathbb{R}.$$
Then there exists a disk K with center 0 such that in each sector of a good k-dissection of the disk there is a representative of a germ that normalizes f and conjugates f to the unit time map of the same vector field, independent of the sector.

REMARK. For any $\alpha \in (\pi/k, 2\pi/k)$ the disk K may be chosen such that the normalizing maps in the theorem will extend to sectors with angle α that cover K and have the same bisectors as the sectors of the good k-dissection. These sectors with angle α at the vertex form a good k-cover of the disk. The normalizing maps may be chosen such that they will all have the same asymptotic Taylor expansions. In this case the overlap function between two normalizing maps defined on intersecting sectors differs from the identity map by an error term that tends to zero exponentially as the vertex of the sector is approached, in fact no slower than $\exp(-C/|z|^k)$ where C is some positive constant depending on the choice of scale: see [2], [11], [14], [15].

The collection of normalizing maps, as described in the Theorem on sectorial normalization, is called a *normalizing cochain*.

8. Correspondence maps for degenerate elementary singularities

The results of this section allow us to give a precise description of the class of such maps. The account follows [6].

Let v be the germ of an analytic vector field at a degenerate elementary singularity of multiplicity $k+1$ at 0 in the plane $\mathbb{C}^2$, and suppose that v is real in $\mathbb{R}^2$. The germ v has an invariant holomorphic curve W^h (h for *hyperbolic*) which is tangent at 0 to that eigenvector of the linear part of v which has nonzero eigenvalue. Let Δ_h be the monodromy map for v corresponding to a positively oriented loop in W_h going once round 0; this map is defined on a transversal Γ that intersects $\mathbb{R}^2$ in a segment containing an "incoming" transversal Γ^+ of a hyperbolic sector of v as in Figure 2. We have
$$\Delta_h(z) = z - 2\pi i z^{k+1} + \cdots.$$

Let F_{norm} be a normalizing cochain for Δ_h, and let f_{norm} be either of the two maps of the collection F_{norm} that are defined on those sectors S_1 and S_2 of the good k-cover that contain a segment of the positive semiaxis. The germ v is formally orbitally equivalent to the germ v_{St} given by a system of the form
$$\dot{z} = z^{k+1}(1 + az^k)^{-1},$$
$$\dot{w} = -w,$$
where the constant a is real and defined uniquely by the germ v.

Let Δ_{St} be the correspondence map to the center manifold for v_{St} (see [5]).

LEMMA. *Under the above hypotheses the correspondence map Δ to the center manifold for v has the form*
$$\Delta = g \circ \Delta_{\text{St}} \circ h_{\text{norm}},$$
where $g: (\mathbb{C}, 0) \to (\mathbb{C}, 0)$ is a holomorphic map germ, and Δ_{St} and f_{norm} are as defined above.

REMARKS. 1. If one of the sectors S_1 and S_2 is exchanged with another, then f_{norm} is altered, but the germ g is also altered and the product Δ remains unchanged.

2. The map Δ_{St} has the form
$$\Delta_{\text{St}} = c \circ f_0 \circ h_{k,a},$$

where

$$c \in \mathbb{R}^+, \qquad f_0: z \mapsto e^{-1/z}, \qquad h_{k,a} = kz^k(1 - az^k \ln z)^{-1}.$$

This concludes the description of correspondence maps for degenerate elementary singularities.

9. The structure of the monodromy map of a separatrix cycle

THEOREM. *After a suitable choice of semitransversal and coordinate chart on it, the monodromy map of a separatrix cycle of an analytic vector field can be expressed as the composition of germs of*
almost regular maps,
normalizing map-cochains and their inverses,
standard flat maps $x \mapsto \exp(-1/x)$ *and their inverses.*

For elementary separatrix cycles the theorem follows immediately from the results of §§5-7. The general case can be reduced to this one by desingularization.

This concludes the first part of the proof of the finiteness theorem, namely the description of the components into which a monodromy map may be decomposed. Finite compositions of such components form a class of maps which is larger than the set of monodromy maps. All these finite compositions are either the identity or have no fixed points apart from zero. The proof of this fact is carried out according to the plan described in §2 and constitutes the most intricate part of the work.

10. Historical remarks

Several variants of the finiteness theorem have been proved in [10, §20], [16], [9], [4], [5], [8], and [6]. Theorems 1-3 were proved in [4] for analytic vector fields with nondegenerate singularities. This implies the results of [9] and [16].

The finiteness theorem for quadratic vector fields was proved by R. Bamon.

THEOREM [8]. *A polynomial vector field on the real plane whose components are polynomials of degree at most 2 have only finitely many limit cycles.*

This implies the results of [16] and [9]. A considerable part of the proof of Bamon's theorem was obtained independently by students M. G. Golitsina and A. Yu. Kotova.

After this manuscript was completed, the author learned of the work [12] in which the finiteness theorem is announced. The author's announcement is in [7].

Bibliography

1. V. I. Arnol'd and Yu. S. Il'yashenko, *Ordinary differential equations*, in: Contemporary Problems of Mathematics, vol. 1, VINITI, Moscow, 1985, pp. 7-149.
2. S. M. Voronin, *Analytic classification of germs of conformal mappings* $(\mathbb{C}, 0) \to (\mathbb{C}, 0)$, Funktsional. Anal. i Prilozhen., **15** (1981), 1-17; English transl. in Functional Anal. Appl. **15** (1981).
3. Yu. S. Il'yashenko, *Singularities and limit cycles of differential equations on the real and complex plane*, Pushchino, Preprint NIVTS Akad. Nauk SSSR, 1982.
4. ____, *Limit cycles of polynomial vector fields with nondegenerate singular points on the real plane*, Funktsional. Anal. i Prilozhen., **18** (1984), 32-42; English transl. in Functional Anal. Appl. **18** (1984).
5. ____, *Dulac's memoir, "On limit cycles" and related questions of the local theory of differential equations*, Uspekhi Mat. Nauk **40** (1985), 41-78; English transl. in Russian Math. Surveys **40** (1985).
6. ____, *Separatrix lunes of analytic vector fields on the plane*, Vestnik Moskov. Univ. No. 4 (1986), 25-31; Engish transl. in Moscow Math. Bull.
7. ____, *The finiteness theorem for limit cycles*, Uspekhi Mat. Nauk **42** (1987), 223; English transl. in Russian Math. Surveys **42** (1987).
8. R. Bamon, *Quadratic vector fields in the plane have a finite number of limit cycles*, Publ. Math. IHES. no. 64 (1986), 111-142.
9. C. Chicone, D. Shafer, *Separatrix and limit cycles of quadratic systems and Dulac's theorem*, Trans. Amer. Math. Soc. **278** (1983), 585-612.
10. H. Dulac, *Sur les cycles limites*, Bull. Soc. Math. France. **51** (1923), 45-188.
11. J. Ecalle, *Sur les fonctions resurgentes*, t. I, II, Orsay, 1981.
12. J. Ecalle, J. Martinet, R. Moussu, and J.-P. Ramis, *Non-accumulation des cycles limites* I, C.R. Acad. Sci. Paris Sér. I Math. **304** (1987), 375-377.
13. Yu. S. Il'yashenko, *The finiteness problem for the limit cycles of polynomial vector fields in the plane, germs of saddle resonant vector fields, and non-Hausdorff Riemann surfaces*. Lecture Notes in Math. No. 1060, Springer-Verlag, Berlin, New York, 1984.
14. B. Malgrange, *Les Traveaux d'Ecalle et de Martinet-Ramis sur les systèmes dynamiques*, Sém. Bourbaki, Novembre, 1981.
15. J. Martinet and J.-P. Ramis, *Problèmes des modules pour des équations différentielle du premier ordre*, Publ. Math. IHES no. 55 (1982), 64-163.
16. I. Sotomajor and R. Paterlini, *Quadratic vector fields with finitely many periodic orbits*, Lecture Notes in Math. No. 1007, Springer-Verlag, Berlin, New York, 1983, pp. 753-766.

Translated by D. R. CHILLINGWORTH

Identities of Associative Algebras

A. R. KEMER

Let $F\langle X\rangle$ be the free associative algebra over a field F of characteristic 0, generated by a countable set of variables X. The elements of the algebra $F\langle X\rangle$ will be called polynomials.

We say that an associative F-algebra A satisfies the identity

$$f(x_1, \ldots, x_n) = 0, \qquad (1)$$

where $f(x_1, \ldots, x_n) \in F\langle X\rangle$, $x_i \in X$, if for arbitrary $a_1, \ldots, a_n \in A$, the equality $f(a_1, \ldots, a_n) = 0$ is satisfied in the algebra. The set of left-hand sides of all identities (1) that are satisfied in A forms an ideal $T[A]$ of the algebra $F\langle X\rangle$, which is called the ideal of identities of the algebra A. The ideal $T[A]$ is a T-ideal (i.e., a completely characteristic ideal) of the algebra $F\langle X\rangle$. The converse is also true: any T-ideal is the ideal of identities of some associative F-algebra A. The algebra A is called a PI-algebra if $T[A] \neq \{0\}$.

A minimal set of polynomials from the T-ideal $T[A]$ that generates $T[A]$ as a T-ideal is called a basis of identities of the algebra A.

The central question in the theory of PI-algebras is Specht's problem (see [1], [2]): Does every associative algebra over a field of characteristic zero have a finite basis of identities?

Specht's problem is part of the problem of classification of T-ideals.

We note two of the most important results on Specht's problem. In 1977 Latyshev [3] and independently G. Genov and A. Popov proved that any associative algebra over a field of characteristic zero, satisfying identities of the form

$$[x_1, \ldots, x_n] \cdot \cdots \cdot [y_1, \ldots, y_n] = 0$$

1980 *Mathematics Subject Classification* (1985 *Revision*). Primary 16A38; Secondary 16A03, 16A46.

has a finite basis of identities. In 1982 A. V. Yakovlev announced the following result: The full algebra of matrices of any order over a field of characteristic zero has a finite basis of identities.

In the present paper Specht's problem is solved completely (positively).

Since finitely generated (f.g.) *PI*-algebras are sufficiently well understood, it is natural to reduce questions about arbitrary *PI*-algebras to (generally speaking, different) questions about f.g. *PI*-algebras. In [4] the theory of identities is reduced to the theory of graded identities of f.g. Z_2-graded algebras. We give the necessary definitions.

An associative algebra A over a field F is called Z_2-graded if A has two distinguished subsets A_0 and A_1 satisfying the conditions:

$$A = A_0 + A_1; \quad A_0^2, A_1^2 \subseteq A_0; \quad A_0 A_1, A_1 A_0 \subseteq A_1.$$

We divide the set of variables X into two disjoint countable subsets Y and Z. We denote by $\mathscr{F}_0$ the subspace of the algebra $F\langle X\rangle$ generated by the monomials in variables from X having even (total) degree in the variables from Z, and by $\mathscr{F}_1$ the subspace generated by monomials of odd degree in Z. An algebra $F\langle X\rangle$ with grading $(\mathscr{F}_0, \mathscr{F}_1)$ is called a free Z_2-graded algebra.

We call an ideal Γ a T_2-ideal if $\varphi(\Gamma) \subseteq \Gamma$ for any endomorphism φ of the algebra $F\langle X\rangle$ such that $\varphi(\mathscr{F}_0) \subseteq \mathscr{F}_0$ and $\varphi(\mathscr{F}_1) \subseteq \mathscr{F}_1$.

Let $f(y_1, \ldots, y_n, z_1, \ldots, z_m) \in F\langle X\rangle$, $y_i \in Y$, $z_j \in Z$. We say that the Z_2-graded algebra A satisfies the graded identity

$$f(y_1, \ldots, y_n, z_1, \ldots, z_m) = 0, \tag{2}$$

if for arbitrary $a_1, \ldots, a_n \in A_0$ and $b_1, \ldots, b_m \in A_1$ the equality

$$f(a_1, \ldots, a_n, b_1, \ldots, b_m) = 0$$

is satisfied in the algebra A.

We denote by $T_2[A]$ the ideal of graded identities of the algebra A, generated in $F\langle X\rangle$ by the left-hand sides of all graded identities (2) that are satisfied by A.

From results of [4] it follows that for a positive solution to Specht's problem it is sufficient to prove the following theorem.

THEOREM 1. *For each n, the set of T_2-ideals of the algebra $F\langle X\rangle$ containing the T-ideal $T[M_n(F)]$, where $M_n(F)$ is the full algebra of matrices of order n, satisfies the ascending chain condition (i.e., each ascending chain of T_2-ideals is stabilized at a finite step).*

If $f \in F\langle X\rangle$, we denote by $\{f\}^{T_2}$ the T_2-ideal generated by f. If f is a multilinear polynomial, we denote by $\{f\}_+^{T_2}$ the T_2-ideal generated by all multilinear polynomials from the ideal $\{f\}^{T_2}$ having degree greater than the degree of f.

Suppose that Theorem 1 is false. Then, since $\operatorname{ch} F = 0$, multilinear polynomials $f_1, f_2, \ldots$, $\deg f_1 < \deg f_2 < \ldots$, exist such that for some n the chain of T_2-ideals

$$T[M_n(F)] \subseteq T[M_n(F)] + \{f_1\}^{T_2} \subseteq T[M_n(F)] + \{f_1\}^{T_2} + \{f_2\}^{T_2} \subseteq \cdots$$

is strictly increasing. We set

$$\Gamma = T[M_n(F)] + \{f_1\}_+^{T_2} + \{f_2\}_+^{T_2} + \cdots.$$

The T_2-ideal Γ satisfies the following conditions:

1) $\Gamma \supseteq T[M_n(F)]$ for some n.

2) There is an infinite sequence of multilinear polynomials $f_1, f_2, \ldots$, $\deg f_1 < \deg f_2 < \cdots$, such that $f_i \notin \Gamma$ but $\{f_i\}_+^{T_2} \subseteq \Gamma$, $i = 1, 2, \ldots$.

An arbitrary T_2-ideal satisfying conditions 1) and 2) will be called a singular counterexample to Specht's problem. The set of all singular counterexamples is denoted by $\mathscr{P}$. We have shown that if Specht's problem has a negative solution, then $\mathscr{P} \neq \varnothing$.

Let A be a finite-dimensional Z_2-graded algebra. We show that $T_2[A] \notin \mathscr{P}$. In fact, since the algebra $A/\operatorname{Rad} A$ (where $\operatorname{Rad} A$ is the radical of the algebra A) has a unit, and $1 \in (A/\operatorname{Rad} A)_0$, we have the equalities

$$A_0 = A_0^2 + (A_0 \cap \operatorname{Rad} A), \qquad A_1 = A_0 A_1 + (A_1 \cap \operatorname{Rad} A).$$

Let f_i be a multilinear polynomial, $f_i \notin T_2[A]$, $\{f_i\}_+^{T_2} \subseteq T_2[A]$. Then using the above equalities we obtain

$$f_i(A_0, \ldots, A_0, A_1, \ldots, A_1) \subseteq f_i(\operatorname{Rad} A, \ldots, \operatorname{Rad} A) \subseteq (\operatorname{Rad} A)^{\deg f_i}.$$

Hence it follows that $\deg f_i < c(A)$, where $c(A)$ is the index of nilpotency of $\operatorname{Rad} A$. Therefore, condition 2) is not fulfilled. We have shown that the set $\mathscr{P}$ has the property:

a) If A is a finite-dimensional Z_2-graded algebra, then $T_2[A] \notin \mathscr{P}$.

In connection with property a) the following questions arises. Let Γ be a T_2-ideal with $\Gamma \supseteq T[M_n(F)]$ for some n. Is it true that Γ is the ideal of graded identities of some finite-dimensional Z_2-graded algebra? This is equivalent to the question about the representability of finitely generated relatively free Z_2-graded PI-algebras by matrices of finite order over some large field.

It is not difficult to show that the set $\mathscr{P}$ also has the property:

b) If A is a finite-dimensional Z_2-graded algebra and $\Gamma \in \mathscr{P}$, then either $\Gamma + T_2[A] \in \mathscr{P}$ or $\Gamma \cap T_2[A] \in \mathscr{P}$.

Finally, the following property of the set $\mathscr{P}$ follows from the definition:

c) If $\Gamma \in \mathscr{P}$, then $\Gamma \supseteq T[M_n(F)]$ for some n.

Thus, for the proof of Theorem 1 it is sufficient to prove the following theorem.

THEOREM 2. *Let $\mathscr{P}$ be an arbitrary set of T_2-ideals of the algebra $F\langle X \rangle$, having properties* a), b), *and* c). *Then $\mathscr{P} = \varnothing$.*

A finite-dimensional Z_2-graded algebra over a field F will be called a classical finite-dimensional graded (c.f.g.) algebra if $A = D + \operatorname{Rad} A$ and $D \cap \operatorname{Rad} A = \{0\}$, where D is a graded subalgebra that is a direct sum of a finite family of simple graded subalgebras of one of the following kinds:

1) $M_n(F)$ with grading $(M_n(F), M_n(F))$,
2) $M_n(F)$ with grading
$$\left(\left(\begin{array}{c|c} M_m(F) & 0 \\ \hline 0 & M_{n-m}(F) \end{array}\right), \left(\begin{array}{c|c} 0 & M_m(F) \\ \hline M_{n-m}(F) & 0 \end{array}\right)\right),$$
where $n \geq m \geq 0$,

3) $M_n(B)$, where $B = 1 \cdot F + c \cdot F$, $c^2 = 1$, with grading $(M_n(F), cM_n(F))$.

If F is an algebraically closed field, then any finite-dimensional algebra is a c.f.g. algebra.

Let A be a c.f.g. algebra. We set $A^* = A$ if A is an algebra with unit; but if A is an algebra without unit, then A^* is A with a unit adjoined. We have the decompositions
$$A = D + \operatorname{Rad} A, \qquad A^* = D^{(*)} + \operatorname{Rad} A,$$
where D and $D^{(*)}$ are semisimple c.f.g. subalgebras with grading (D_0, D_1) and $(D_0^{(*)}, D_1^{(*)})$ respectively, and
$$D_0^{(*)} = \bigoplus_{i=1}^{t} M_{n_i}(F).$$

We set
$$a(A) = \sum_{i=1}^{t} n_i,$$
$$b_0(A) = \dim_F D_0, \qquad b_1(A) = \dim_F D_1,$$
and we put $c(A)$ equal to the index of nilpotency of $\operatorname{Rad} A$.

Let $f = f(y_1, \ldots, y_p, z_1, \ldots, z_q)$ be a multilinear polynomial and Λ be a finite ordered set of variables. We introduce the notation

$$S_\Lambda(f) = \begin{cases} \displaystyle\sum_{\sigma \in S(m)} (-1)^\sigma f(y_{\sigma(1)}, \ldots, y_{\sigma(m)}, y_{n+1}, \ldots, y_p, z_1, \ldots, z_q) \\ \qquad\qquad\qquad\qquad\qquad\qquad \text{if } \Lambda = (y_1, \ldots, y_m), \\ \displaystyle\sum_{\sigma \in S(m)} (-1)^\sigma f(y_1, \ldots, y_p, z_{\sigma(1)}, \ldots, z_{\sigma(m)}, z_{m+1}, \ldots, z_q) \\ \qquad\qquad\qquad\qquad\qquad\qquad \text{if } \Lambda = (z_1, \ldots, z_m), \\ 0 \qquad \text{in all remaining cases,} \end{cases}$$

where $S(m)$ is the group of permutations of the set $\{1, \ldots, m\}$; $(-1)^\sigma = 1$, if σ is an even permutation, and $(-1)^\sigma = -1$, if σ is odd.

If A is a c.f.g. algebra and Γ is a T_2-ideal, then we denote by $S^k_{b_0(A), b_1(A)}(\Gamma)$ the T_2-ideal generated by all polynomials of the form $S_{\Lambda_1} \cdots S_{\Lambda_k}(f)$, where $f \in \Gamma$ and $\Lambda_1, \ldots, \Lambda_k$ are disjoint sets of variables such that, for each i, either $\Lambda_i \subseteq Y$ and $|\Lambda_i| = b_0(A) + 1$, or $\Lambda_i \subseteq Z$ and $|\Lambda_i| = b_1(A) + 1$.

It is easy to show that $S^k_{b_0(A), b_1(A)}(\Gamma) \subseteq T_2[A]$ for $k \geq c(A)$. We denote by $d(\Gamma, A)$ the smallest k such that $S^k_{b_0(A), b_1(A)}(\Gamma) = T_2[A]$. It follows from what was said above that

$$d(\Gamma, A) \leq c(A). \qquad (3)$$

Let Γ be a T_2-ideal and A a c.f.g. algebra with $\Gamma \supseteq T_2[A]$. We call the ordered set of numbers

$$t(\Gamma, A) = (a(A), b_0(A), b_1(A), d(\Gamma, A))$$

the type of the pair (Γ, A). We order the types of pairs lexicographically.

Briefly, the idea of the proof of Theorem 2 consists in the following. If $\Gamma \in \mathscr{P}$ and A is a c.f.g. algebra with $\Gamma \supseteq T_2[A]$, then it is necessary to find a pair (Γ', A') such that $\Gamma' \in \mathscr{P}$ and $t(\Gamma', A') < t(\Gamma, A)$. For the realization of this idea, Propositions 1, 2, and 3 are applied.

PROPOSITION 1. *Let A and B be c.f.g. algebras with $T_2[B] \supseteq T_2[A]$. Then c.f.g. subalgebras $C^{(1)}, \ldots, C^{(\mu)}$ of B exist such that $a(C^{(i)}) \leq a(A)$, $b_0(C^{(i)}) \leq b_0(A)$, $b_1(C^{(i)}) \leq b_1(A)$ for each i, and $\prod_{i=1}^{\mu} T_2[C^{(i)}] = T_2[B]$.*

PROPOSITION 2. *Let A be a c.f.g. algebra and Γ a T_2-ideal, $\Gamma \supseteq T_2[A]$, and $d(\Gamma, A) < c(A)$. Then a c.f.g. algebra $A^{(0)}$ and a finite family $\{A^{(i)}\}_{i \in I}$ of c.f.g. subalgebras of A exist such that:*

1) $a(A^{(i)}) < a(A)$, $i \in I$,
2) $a(A^{(0)}) = a(A)$, $b_0(A^{(0)}) = b_0(A)$, $b_1(A^{(0)}) = b_1(A)$, $c(A^{(0)}) \leq d(\Gamma, A)$,
3) $\Gamma \cap T_2[A^{(0)}] \cap (\bigcap_{i \in I} T_2[A^{(i)}]) = T_2[A]$.

PROPOSITION 3. *Let A be a c.f.g. algebra and Γ a T_2-ideal, $\Gamma \supseteq T_2[A]$, and $d(\Gamma, A) = c(A)$. Then a c.f.g. algebra B and a finite family $\{B^{(i)}\}_{i \in I}$ of c.f.g. subalgebras of B exist such that:*

1) $\Gamma \supseteq T_2[B] \supseteq T_2[A]$,
2) $a(B^{(i)}) < a(A)$, $i \in I$,
3) $S^{d(\Gamma, A)-1}_{b_0(A), b_1(A)}\left(\Gamma \cap (\bigcap_{i \in I} T_2[B^{(i)}])\right) \subseteq T_2[B]$.

We show how Theorem 2 follows from these propositions.

From b) it is easy to deduce the following property of $\mathscr{P}$:

d) If $\Gamma \in \mathscr{P}$ and $C^{(1)}, \ldots, C^{(m)}$ are c.f.g. algebras, then either $\Gamma \cap \left(\bigcap_{i=1}^{m} T_2[C^{(i)}]\right) \in \mathscr{P}$ or a T_2-ideal $\Gamma' \in \mathscr{P}$ exists such that

$$\Gamma + T_2[C^{(i)}] \supseteq \Gamma' \supseteq T_2[C^{(i)}] \qquad \text{for some } i.$$

We denote by $\mathscr{B}$ the class of pairs (Γ, A) with $\Gamma \in \mathscr{P}$, A a c.f.g. algebra and $\Gamma \supseteq T_2[A]$. If $\mathscr{P} \neq \varnothing$, then it follows from property c) that $\mathscr{B} \neq \varnothing$.

We select from $\mathscr{B}$ a pair (Γ, A) of minimal type. By virtue of (3), the pair (Γ, A) satisfies the conditions of Proposition 2 or of Proposition 3. If (Γ, A) satisfies the conditions of Proposition 2, then a c.f.g. algebra $A^{(0)}$ and a finite family of c.f.g. algebras $\{A^{(i)}\}_{i \in I}$ exist for which the conclusions of Proposition 2 hold. Hence, since $T_2[A] \notin \mathscr{P}$ and $a(A^{(i)}) < a(A)$, $i \in I$, by property d) a T_2-ideal $\Gamma' \in \mathscr{P}$ exists with $\Gamma' \supseteq T_2[A^{(0)}]$. From (3) and the inequality $c(A^{(0)}) \leq d(\Gamma, A)$ it follows that $d(\Gamma', A^{(0)}) \leq d(\Gamma, A)$, whence from the minimality of $t(\Gamma, A)$ we obtain the equality $d(\Gamma', A^{(0)}) = c(A^{(0)})$. We have shown that among the pairs of minimal type in the class $\mathscr{B}$ there is a pair satisfying the conditions of Proposition 3.

Now let the pair (Γ, A) satisfy the conditions of Proposition 3. Then a c.f.g. algebra B and a finite family of c.f.g. subalgebras $\{B^{(i)}\}_{i \in I}$ of B exist such that

$$a(B^{(i)}) < a(A), \quad \Gamma \supseteq T_2[B] \supseteq T_2[A], \quad S_{b_0(A), b_1(A)}^{d(\Gamma, A)-1}(\Gamma_1) \subseteq T_2[B], \qquad (4)$$

where $\Gamma' = \Gamma \cap (\bigcap_{i \in I} T_2[B^{(i)}])$.

Since $a(B^{(i)}) < a(A)$, by property d) we have $(\Gamma_1, B) \in \mathscr{B}$. By Proposition 1 there are c.f.g. algebras $C^{(1)}, \ldots, C^{(\mu)}$ satisfying

$$a(C^{(i)}) \leq a(A), \quad b_0(C^{(i)}) \leq b_0(A), \quad b_1(C^{(i)}) \leq b_1(A), \qquad (5)$$

$$i = 1, \ldots, \mu, \quad \bigcap_{i=1}^{\mu} T_2[C^{(i)}] = T_2[B].$$

Hence, since $\Gamma_1 \supseteq T_2[B]$, we obtain the equality

$$\Gamma_1 \cap \left(\bigcap_{i=1}^{\mu} T_2[C^{(i)}]\right) = T_2[B].$$

Since $T_2[B] \notin \mathscr{P}$, by d) there is a T_2-ideal $\Gamma_2 \in \mathscr{P}$ such that

$$\Gamma_1 + T_2[C^{(k)}] \supseteq \Gamma_2 \supseteq T_2[C^{(k)}] \qquad (6)$$

for some k; in particular, $(\Gamma_2, C^{(k)}) \in \mathscr{B}$. We shall show that $t(\Gamma_2, C^{(k)}) < t(\Gamma, A)$. This will contradict the minimality of the type $t(\Gamma, A)$.

If even one of the inequalities (5) (for $i = k$) is strict, then $t(\Gamma_2, C^{(k)}) < t(\Gamma, A)$. In the contrary case, from (4) and (6) we obtain

$$S^{d(\Gamma, A)-1}_{b_0(C^{(k)}), b_1(C^{(k)})}(\Gamma_2) = S^{d(\Gamma, A)-1}_{b_0(A), b_1(A)}(\Gamma_2) \subseteq T_2[C^{(k)}],$$

i.e., $d(\Gamma_2, C^{(k)}) < d(\Gamma, A)$.

Bibliography

1. Wilhelm Specht, Gesetze in Ringen, Math. Z. **52** (1950), 557–589.
2. The Dniester notebook: Unsolved problems in the theory of rings and modules, 2nd ed., Inst. Mat. Sibirsk. Otdel. Akad. Nauk SSSR, Novosibirsk, 1976. (Russian)
3. V. N. Latyshev, *Nonmatrix varieties of associative algebras*, Doctoral Dissertation, Moscow, 1978. (Russian)
4. A. R. Kemer, *Varieties and Z_2-graded algebras*, Izv. Akad. Nauk SSSR Ser. Mat. **48** (1984), 1042–1059; English transl. in Math. USSR Izv. **25** (1985).

Translated by C. W. KOHLS

Lie Groups and Ergodic Theory

UDC 512.817+512.626.7

G. A. MARGULIS

In these notes we present some interconnections between the theory of Lie groups and ergodic theory. Some applications to number theory will also be given. Since the author addresses himself mostly to algebraists, some space is devoted to a survey of elementary facts from ergodic theory.

As usual, $\mathbb{R}$, $\mathbb{Q}$, $\mathbb{Z}$ and $\mathbb{N}^+$ denote the sets of real numbers, rational numbers, integers and positive integers, respectively.

1. Preliminary facts from ergodic theory

1.1. Ergodic theory is the study of groups (and more generally, semigroups) of transformations of measure spaces. By a *measure* on a space X we understand a countably additive, finite- or infinite-valued set function μ defined on some σ-algebra Ω of subsets of X. Subsets lying in Ω are called μ-*measurable* or simply *measurable*. A measure μ on X is called *normalized* if $\mu(X) = 1$, *finite* if $\mu(X) < \infty$, and σ-*finite* if X can be written as a union of a countable number of measurable sets of finite measure. In what follows, all measures are assumed to be σ-finite.

By a *Borel measure* on a locally compact topological space X we understand a measure defined on the σ-algebra of the Borel subsets. A Borel measure is called *locally finite* if the measure of every compact subset is finite. If X is a separable metric space, then every locally finite Borel measure on X is σ-finite and *regular* (i.e. the measure of every closed set is the infimum of the measures of the open sets containing it).

In measure theory, the expressions "almost everywhere" and "for almost every x" mean "everywhere, except for a set of measure zero" and "for every x outside some set of measure zero".

1.2. Let (X, μ) be a measure space. A mapping of X into a measure

---

1980 *Mathematics Subject Classification* (1985 *Revision*). Primary 22E45, 22D40, 28D05; Secondary 11Q15.

space is called *measurable* if the inverse image of every measurable set is measurable. A measurable mapping $T: X \to X$ is called an *endomorphism* of X if it preserves μ, i.e. if $\mu(A) = \mu(T^{-1}A)$ for every measurable $A \subset X$. An endomorphism of X is called an *automorphism* if it is a one-to-one mapping.

Let G be a locally compact group. We say that G *acts on* X if $gx \in X$ is defined for every $g \in G$ and $x \in X$, and the following conditions are satisfied:

(a) $g_1(g_2 x) = (g_1 g_2)x$ and $ex = x$ for all $g_1, g_2 \in G$ and $x \in X$;

(b) the mapping $(g, x) \mapsto gx$ is measurable, i.e. the set $\{(g, x) \in G \times X | gx \in A\}$ belongs to the natural σ-algebra of subsets of $G \times X$, for every measurable set $A \subset X$;

(c) μ is G-*invariant*, i.e. $\mu(gA) = \mu(A)$ for every $g \in G$ and every measurable set $A \subset X$ (or, in other words, the mapping $x \mapsto gx$, $x \in X$, is an automorphism of X, for every $g \in G$).

If G acts on X, then we say that X is a G-space. In G-spaces, G acts on X on the left. For this reason, G-spaces are also called *left G-spaces*. Sometimes it is convenient to consider *right G-spaces*, when xg is defined for all $g \in G$ and $x \in X$, and conditions analogous to (a), (b), and (c) are satisfied.

We say that a G-space X is *ergodic*, or that the action of G on X is *ergodic*, if the following condition is satisfied: whenever $A \subset X$ is measurable and $\mu(A \triangle (gA)) = 0$ for all $g \in G$, then either $\mu(A) = 0$ or $\mu(X - A) = 0$. In this case, we also say that μ is G-*ergodic* or simply *ergodic*. Ergodicity is similarly defined for right G-spaces. For σ-compact G, ergodicity of the action of G on X is equivalent to the following condition: whenever $A \subset X$ is measurable and $GA = A$, then either $\mu(A) = 0$ or $\mu(X - A) = 0$. Hence, if G is σ-compact and G acts transitively on X, then the action of G on X is ergodic (this assertion is, in general, false for arbitrary G).

An automorphism T of a space X is called *ergodic* if the action of the group $\{T^n | n \in \mathbb{Z}\}$ on X is ergodic. It is easy to see that ergodicity of T is equivalent to the following property: whenever $A \subset X$ is measurable and $T(A) = A$, then either $\mu(A) = 0$ or $\mu(X - A) = 0$.

The study of arbitrary G-spaces reduces, in a well-known sense, to the study of ergodic G-spaces. Namely, under sufficiently general conditions (in particular, when X is a locally compact separable metric space, μ is a locally finite Borel measure and G is metrizable and separable), one can find a partition of X into G-invariant sets X_y, $y \in Y$, measures μ_y on X_y and a measure ν on Y such that: 1) the set $A \cap X_y$ is μ_y-measurable for all measurable $A \subset X$ and almost all $y \in Y$, and

$$\mu(A) = \int_Y \mu_y(A \cap X_y) \, d\nu(y);$$

2) the restriction of the group action to X_y is ergodic with respect to μ_y for almost all $y \in Y$.

When μ is finite, we say that an automorphism T is *mixing* if for all measurable $A, B \subset X$
$$\lim_{n \to \infty} \tilde{\mu}((T^n A) \cap B) = \tilde{\mu}(A)\tilde{\mu}(B),$$
where $\tilde{\mu}(C) = \mu(C)/\mu(X)$, $C \subset X$. It is easy to see that if T is mixing, then T is ergodic. However, the converse is, in general, false.

1.3. Suppose that the locally compact group G acts on the measure space (X, μ). We denote by $L_p(X, \mu)$ the space of complex-valued, p-integrable functions on X, and by $\| \ \|_p$ the norm on $L_p(X, \mu)$. If $p \geq 1$, we define the representation ρ of G on $L_p(X, \mu)$ by putting
$$(\rho(g)f)(x) = f(g^{-1}x), \qquad f \in L_p(X, \mu), \qquad g \in G.$$
Since μ is G-invariant, $\|\rho(g)f\|_p = \|f\|_p$ for all $g \in G$ and $f \in L_p(X, \mu)$. In particular, ρ defines a unitary representation of G on $L_2(X, \mu)$. It is not difficult to show that this representation is continuous, i.e. that $\rho(g)f$ depends continuously on $(g, f) \in G \times L_2(X, \mu)$. The following statements are easy to prove:

(1) Suppose that $p \geq 1$, and μ is finite. Then the G-space X is ergodic if and only if every $\rho(G)$-invariant function $f \in L_p(X, \mu)$ is constant.

(2) Denote by $L_2^0(X, \mu)$ the subspace of $L_2(X, \mu)$ consisting of those functions whose integral is zero. Let $g \in G$ and $T(x) = gx$, $x \in X$. If μ is finite and the unitary operator $\rho(g)$ restricted to $L_2^0(X, \mu)$ has absolutely continuous spectrum, then T is mixing.

1.4. The following simple but important theorem holds.

POINCARÉ RECURRENCE THEOREM. *Let T be an endomorphism of the measure space (X, μ). Suppose that μ is finite. Then for every measurable $A \subset X$ and μ-almost every $x \in A$, the set $\{n \in \mathbb{N}^+ | T^n x \in A\}$ is infinite.*

SKETCH OF PROOF. Let $B_m = \{x \in A | T^n x \notin A \text{ for all } n \geq m\}$, where $m \in \mathbb{N}^+$. Then $T^{-mi} B_m \cap T^{-mj} B_m = \varnothing$ for all $i, j \in \mathbb{N}^+$, $i \neq j$. But T preserves μ, and μ is finite. Hence $\mu(B_m) = 0$ and it follows that the μ-measure of the set $\{x \in A | \text{ the set } \{n \in \mathbb{N}^+ | T^n x \in A\} \text{ is finite}\} = \bigcup_{m \in \mathbb{N}^+} B_m$ is equal to zero.

An analog of the Poincaré recurrence theorem for G-spaces is the following:

THEOREM. *Let G be a locally compact, σ-compact group and let X be a G-space with finite measure μ. Then for every measurable $A \subset X$ and almost every $x \in A$, the set $\{g \in G | gx \in A\}$ is not relatively compact in G.*

1.5. A central result of ergodic theory is

BIRKHOFF'S INDIVIDUAL ERGODIC THEOREM. *Suppose that T is an automorphism of the space X with finite measure μ, and let $f \in L_1(X, \mu)$. Let*

$$f_n(x) = \frac{1}{n} \sum_{i=1}^{n} f(T^i x), \qquad \text{if} \quad n > 0,$$

$$f_n(x) = \frac{1}{-n} \sum_{i=1}^{n} f(T^{-i} x), \qquad \text{if} \quad n < 0.$$

Then there exists a function $\tilde{f} \in L_1(X, \mu)$ such that $\tilde{f}(Tx) = \tilde{f}(x)$, $\int_X \tilde{f}(x) d\mu(x) = \int_X f(x) d\mu(x)$ and $\lim_{n\to\infty} f_n(x) = \tilde{f}(x)$ for almost every $x \in X$. If the automorphism T is ergodic, then $\tilde{f}(x) = (1/\mu(x)) \int_X f(x) d\mu(x)$.

The proof of this theorem, which can be found in any textbook on ergodic theory (see, for example, [2], [11]), is rather complicated. The so-called mean ergodic theorem, which asserts that f_n tends to f in the norm of $L_1(X, \mu)$, is simpler to prove. There are analogs of the individual and mean ergodic theorems for G-spaces (see [20]); for this, we replace f_n by functions of the form

$$f_n(x) = \frac{1}{\mu_G(K_n)} \int f(g^{-1} x) d\mu_G(g),$$

where $\{K_n\}$ is a suitable exhaustive sequence of compact sets in G, and μ_G is the left-invariant Haar measure on G. It should be noted that a generalization of this sort, far from being always possible, is valid only for certain classes of groups G (in particular, for compactly generated nilpotent groups).

1.6. Let now X be a locally compact separable metric space, G a separable metrizable locally compact group, and consider a continuous action of G on X (i.e. $gx \in X$ is defined for all $g \in G$ and $x \in X$, and (a) $g_1(g_2 x) = (g_1 g_2)x$ and $ex = x$; (b) gx depends continuously on $(g, x) \in G \times X$). Denote by Ω the set of non-trivial G-invariant locally finite Borel measures on X, and by Ω_0 the set of G-ergodic measures $\mu \in \Omega$. The set Ω forms a cone in the space of measures and Ω_0 coincides with the set of extreme points of this cone ($\mu \in \Omega$ is called *extreme* if $\mu = \mu_1 + \mu_2$, where $\mu_1 \in \Omega$ and $\mu_2 \in \Omega$, implies that μ_1 and μ_2 are multiples of μ). An arbitrary measure $\mu \in \Omega$ decomposes into an integral of measures from Ω_0, i.e. there is a measure ν_μ on Ω_0 such that

$$\mu(A) = \int_{\Omega_0} \omega(A) d\nu_\mu(\omega)$$

for every μ-measurable subset $A \subset X$.

If G is amenable, and X is compact, then $\Omega \neq \varnothing$ (a group G is called *amenable* if there exists an invariant mean on the space of bounded continuous functions on G; every solvable group is amenable). This property characterizes amenable groups, i.e. if G is not amenable, then there exists a continuous action of G on some compact metric space Y such that there is no nontrivial, finite, G-invariant Borel measure on Y.

We say that the action of G on X is *strictly ergodic* if $\Omega = \Omega_0$ or, in other words, any two measures in Ω_0 are multiples of each other. We shall call a homeomorphism $T: X \to X$ strictly ergodic if the action of the group $\{T^n | n \in \mathbb{Z}\}$ on X is strictly ergodic. If X is compact, then a homeomorphism T is strictly ergodic if and only if given any continuous function f on X, there is a constant $c(f)$ such that $\lim f_n(x) = c(f)$ for every $x \in X$, where the f_n's are defined in the formulation of Birkhoff's theorem (in addition, a) $c(f) = \int_X f(x) d\mu(x)$, where μ is the unique normalized T-invariant Borel measure on X; b) $f_n(x)$ tends to $c(f)$ uniformly in x).

We call a measure μ on X *strictly positive* if $\mu(U) > 0$ for every nonempty open $U \subset X$. If $\mu \in \Omega$ and μ is strictly positive, then it follows from the G-ergodicity of μ that the orbit Gx is dense in X for μ-almost every $x \in X$ (the converse is, in general, false). If X is compact, the homeomorphism $T: X \to X$ is strictly ergodic, and the unique normalized T-invariant Borel measure on X is strictly positive, then the orbit $\{T^n x | n \in \mathbb{Z}\}$ is dense in X for every $x \in X$ (however, there exists non-strictly ergodic homeomorphisms $T: X \to X$ of compact spaces such that the orbit $\{T^n x | n \in \mathbb{Z}\}$ is dense in X for every $x \in X$).

2. Ergodic properties of actions on homogeneous spaces

2.1. Results from the theory of infinite dimensional, unitary representations of groups play an essential role in the study of ergodic properties of actions on homogeneous spaces. Some of these results will be mentioned in §§2.2–2.4 and 2.7–2.9.

2.2. GENERALIZED MAUTNER LEMMA. *Let H be a topological group, and x and y elements of H such that the sequence $\{x^n y x^{-n}\}$ converges to e as $n \to \infty$. If ρ is a continuous unitary representation of H in a Hilbert space W, $w \in W$, and $\rho(x)w = w$, then $\rho(y)w = w$.*

PROOF. Since $\rho(x)w = w$, and ρ is unitary, we have
$$\|\rho(y)w - w\| = \|\rho(y)\rho(x^{-n})w - \rho(x^{-n})w\| = \|\rho(x^n y x^{-n})w - w\|$$
for all $n \in \mathbb{Z}$. But $\{x^n y x^{-n}\}$ converges to e as $n \to \infty$, and ρ is continuous. Therefore, $\|\rho(y)w - w\| = 0$ and hence $\rho(y)w = w$.

2.3. COROLLARY. *Let ρ be a continuous unitary representation of the group $SL_2(\mathbb{R})$ in a Hilbert space W, $w \in W$ and $d = \begin{pmatrix} a & 0 \\ 0 & a^{-1} \end{pmatrix} \in SL_2(\mathbb{R})$, where $a \neq \pm 1$. If $\rho(d)w = w$, then $\rho(SL_2(\mathbb{R}))w = w$.*

PROOF. Let
$$U = \left\{ \begin{pmatrix} 1 & x \\ 0 & 1 \end{pmatrix} \Big| x \in \mathbb{R} \right\} \quad \text{and} \quad U^- = \left\{ \begin{pmatrix} 1 & 0 \\ x & 1 \end{pmatrix} \Big| x \in \mathbb{R} \right\}.$$

By replacing d with d^{-1}, if necessary, we may assume that $|a| < 1$. Then, as is easy to check, the sequence $\{d^n u d^{-n}\}$ converges to e as $n \to \infty$, for every $u \in U$. It follows from this and Lemma 2.2 that $\rho(U)w = w$. Similarly, by replacing d and U with d^{-1} and U^-, we get that $\rho(U^-)w = w$. But the subgroups U and U^- generate $SL_2(\mathbb{R})$. Hence $\rho(SL_2(\mathbb{R}))w = w$.

2.4. PROPOSITION. *Let ρ be a continuous unitary representation of $SL_2(\mathbb{R})$ in a Hilbert space W. Let $U = \{\begin{pmatrix} 1 & x \\ 0 & 1 \end{pmatrix} | x \in \mathbb{R}\}$. If $w \in W$ is such that $\rho(U)w = w$, then $\rho(SL_2(\mathbb{R}))w = w$.*

PROOF. We may assume that $\|w\| = 1$. Consider the continuous function $\varphi(g) = \langle \rho(g), w, w \rangle$, where $g \in SL_2(\mathbb{R})$ and $\langle \, , \, \rangle$ denotes the inner product. Since ρ is unitary, and $\rho(U)w = w$, φ is constant on the two-sided cosets of U. But if $\begin{pmatrix} a & b \\ c & d \end{pmatrix} \in SL_2(\mathbb{R})$ and $c \neq 0$, then

$$\begin{pmatrix} 1 & c^{-1}(1-a) \\ 0 & 1 \end{pmatrix} \begin{pmatrix} a & b \\ c & d \end{pmatrix} \begin{pmatrix} 1 & c^{-1}(1-d) \\ 0 & 1 \end{pmatrix} = \begin{pmatrix} 1 & 0 \\ c & 1 \end{pmatrix}.$$

Therefore, when $c \neq 0$,

$$\varphi\left(\begin{pmatrix} a & 0 \\ c & a^{-1} \end{pmatrix}\right) = \varphi\left(\begin{pmatrix} 1 & 0 \\ c & 1 \end{pmatrix}\right).$$

Letting $c \to 0$ in this equation, we get

$$\varphi\left(\begin{pmatrix} a & 0 \\ 0 & a^{-1} \end{pmatrix}\right) = 1.$$

From this and the unitarity of ρ, it follows that $\rho(g)w = w$ for all $g = \begin{pmatrix} a & 0 \\ 0 & a^{-1} \end{pmatrix}$. Using Corollary 2.3, we not obtain that $\rho(SL_2(\mathbb{R}))w = w$.

2.5. Let G be a locally compact, σ-compact group. We denote by μ_G the right-invariant Haar measure on G. Given any discrete subgroup $\Gamma \subset G$, μ_G induces a measure on G/Γ, which we also denote by μ_G. A discrete subgroup $\Gamma \subset G$ is called a *lattice* if $\mu_G(G/\Gamma) < \infty$. A lattice Γ is called *uniform* if G/Γ is compact, and nonuniform otherwise; if G contains a lattice, then G is *unimodular*, i.e. μ_G is left-invariant.

G acts on G/Γ by left translations, and μ_G is invariant under this action if and only if G is unimodular.

LEMMA. *Let G be a unimodular group. Given any discrete subgroup $\Gamma \subset G$, the G-invariant measure μ_G on G/Γ is G-ergodic.*

LEMMA. *Given any discrete subgroup $\Gamma \subset G$, the measure μ_G on G/Γ is G-ergodic.*

In view of this lemma and (1) of §1.3, Corollary 2.3 and Proposition 2.4 imply the following:

2.6. THEOREM. *Let $G = SL_2(\mathbb{R})$ and Γ a lattice in G.*

(a) *Let $d = \begin{pmatrix} a & 0 \\ 0 & a^{-1} \end{pmatrix} \in G$, where $a \neq \pm 1$. Then the automorphism $x \mapsto dx$ of G/Γ (with measure μ_G), where $x \in G/\Gamma$, is ergodic.*

(b) Let $U = \{ \begin{pmatrix} 1 & x \\ 0 & 1 \end{pmatrix} | x \in \mathbb{R} \}$. *Then the action of U on G/Γ by left translations is ergodic.*

2.7. Corollary 2.3 and Proposition 2.4 are a special case of Theorem 2.9 below. Before formulating this theorem, we give the following

DEFINITION. Let G be a topological group, $H \subset G$ and $F \subset G$. We say that the triple (G, H, F) has the *Mautner property* if the following condition is satisfied: $\rho(H)w = w$ for every continuous unitary representation ρ of G in a Hilbert space W and every $w \in W$ such that $\rho(F)w = w$.

Lemma 2.2, Corollary 2.3, and Proposition 2.4 assert that the triples $(H, \{y\}, \{x\})$, $(SL_2(\mathbb{R}), SL_2(\mathbb{R}), \{d\})$, and $(SL_2(\mathbb{R}), SL_2(\mathbb{R}), U)$, respectively, have the Mautner property.

2.8. Let G be a connected Lie group, $\mathfrak{g}$ its Lie algebra, and Ad the adjoint representation of G. We say that the subgroup $F \subset G$ is Ad-*compact* if the subgroup $\text{Ad}(F)$ is relatively compact in the group of linear transformations of $\mathfrak{g}$. If H_1 and H_2 are two connected, normal subgroups of G such that the images of F in G/H_1 and G/H_2 are Ad-compact, then the image of F in $G/(H_1 \cap H_2)^0$ is also Ad-compact, where $(H_1 \cap H_2)^0$ is the connected component of $H_1 \cap H_2$ containing the identity. Hence there exists a unique minimal connected normal subgroup H_F of G such that the image of F in G/H_F is Ad-compact. If G is a connected semisimple Lie group with trivial center, then H_F coincides with the product of those simple factors G_i of G for which the subgroup $\pi_i(F)$ is not relatively compact in G_i, where $\pi_i: G \to G_i$ is the natural projection.

2.9. THEOREM (see [3], [16]). *Let G be a connected Lie group, F a subgroup of G, and let H_F be as in §2.8. Suppose that either G is semisimple, or F is a one-parameter subgroup. Then the triple (G, H_F, F) has the Mautner property.*

2.10. Using Theorem 2.9, it is not difficult to obtain Theorem 2.11, formulated below, on the ergodicity of actions on homogeneous spaces. To this end, we need only use (1) of §1.3 and the following simple lemma:

LEMMA. *Let G be a unimodular, locally compact, σ-compact group, H a normal subgroup of G, and $\Gamma \subset G$ a discrete subgroup. Suppose that the subgroup $H \cdot \Gamma$ is dense in G. Then the action of H on G/Γ by left translations is ergodic.*

2.11. THEOREM (see [13]). *Let G be a connected Lie group, Γ a lattice in G, F a subgroup of G, and let H_F be as in §2.8. Suppose that the following conditions are satisfied: (a) either G is semisimple, or F is a one-parameter subgroup; (b) the subgroup $H_F \cdot \Gamma$ is dense in G. Then the action of F on G/Γ by left translations is ergodic.*

2.12. We mention a result on the mixing property for actions on homogeneous spaces.

THEOREM (see [3]). *Let G be a connected semisimple Lie group, Γ a lattice in G, $F \subset G$ a one-parameter subgroup, and let H_F be as in §2.8. Suppose that the subgroup $H_F \cdot \Gamma$ is dense in G. Then given $g \in F$, $g \neq e$, the automorphism $x \mapsto gx$ of G/Γ with measure μ_G, where $x \in G/\Gamma$, is mixing.*

This theorem is easily derived from (1) and (2) of §1.3, Lemma 2.10, and the following result from representation theory:

2.13. THEOREM (see [16]). *Let G be a connected semisimple Lie group, $F \subset G$ a one-parameter subgroup, H_F as in §2.8, and let ρ be a continuous unitary representation of G in a Hilbert space W. Suppose that there are no nonzero $\rho(H_F)$-invariant vectors in W. Then the operator $\rho(g)$ has absolutely continuous spectrum for all $g \in F$, $g \neq e$.*

3. Orbit closures and invariant measures for actions on homogeneous spaces

3.1. Some of the results of the previous section can be interpreted as statements about the behavior of typical orbits. In particular, it follows from these results that, under sufficiently general conditions, "almost every" orbit is dense in a homogeneous space. However, an "individual" orbit can, in general, be arranged to be arbitrarily complicated. In fact, if $G = SL_2(\mathbb{R})$, Γ is a lattice in G and $\mathscr{D} \subset G$ is the cyclic subgroup generated by a nontrivial diagonal matrix, then there exists an $x \in G/\Gamma$ such that the closure of the orbit $\mathscr{D}x$ in G/Γ is a Cantor set. However, if instead of $\mathscr{D}$ we take a unipotent subgroup, then the situation is quite different. Specifically, the following holds:

3.2. THEOREM (see [5]). *Let $G = SL_2(\mathbb{R})$, $u = \begin{pmatrix} 1 & 1 \\ 0 & 1 \end{pmatrix}$, and let Γ be a uniform lattice in G. Then the homeomorphism $x \mapsto ux$ of G/Γ, where $x \in G/\Gamma$, is strictly ergodic, and, therefore, the orbit $\{u^n x \mid n \in \mathbb{Z}\}$ is dense in G/Γ, for every $x \in G/\Gamma$.*

3.3. In view of Theorem 3.2, it might be conjectured that if G is a connected semisimple Lie group and Γ a uniform lattice in G, then, under some natural restrictions on G and Γ, the action of the unipotent subgroups on G/Γ by left translations are strictly ergodic. However, this is not so. In particular, there is, even in $G = SL_3(\mathbb{R})$, a uniform lattice Γ and a one-parameter unipotent subgroup U such that, for some $x \in G/\Gamma$, the closure of the orbit Ux is a submanifold of positive codimension. Nonetheless, the following two conjectures are very plausible.

CONJECTURE 1. Let G be a connected Lie group, Γ a lattice in G, and U a subgroup of G. Suppose that U is nilpotent, i.e. the transformation $\operatorname{Ad} u$ is unipotent for every $u \in U$. Then, given any U-invariant, U-ergodic, locally finite Borel measure σ on G/Γ, there exist a closed subgroup $P \subset G$

containing U and $x \in X$ such that the set Px is closed in G/Γ and σ is a finite P-invariant measure whose support coincides with Px.

CONJECTURE 2. *Suppose, as in Conjecture 1, that G is a connected Lie group, Γ a lattice in G, and U a unipotent subgroup of G. Then, given any point $x \in G/\Gamma$, there exists a closed subgroup $P \subset G$ containing U such that the closure of the orbit Ux coincides with Px.*

Note that if the orbit Px is closed, then the natural mapping $P/(P \cap G_x) \mapsto Px$ is a homeomorphism, where $G_x = \{g \in G | gx = x\}$ is the stabilizer subgroup of x.

Conjectures 1 and 2 were formulated in [4] (when G is reductive and U is a one-parameter subgroup). In the same paper it is remarked that Conjecture 2 is due to Ragunathan, who also investigated the connection of his conjecture with Davenport's conjecture (see Theorem 4.4 of this paper). Since the group U is nilpotent, for every action of this group on some compact space X there exists a finite U-invariant Borel measure on X. Therefore, when the closure $\overline{Ux}$ of the orbit Ux is a compact U-minimal set, the validity of Conjecture 2 follows from the validity of Conjecture 1 (U-minimal means that the orbit Uy is dense in $\overline{Ux}$ for every $y \in \overline{Ux}$).

Conjecture 1 is proved in [4] for reductive G and for maximal unipotent subgroups U, and Conjecture 2 is proved in [7] for reductive G and for horospherical U. (A subgroup $\mathscr{W}$ is called *horospherical* if there exists $g \in G$ such that $\mathscr{W} = \{w \in G | g^j w g^{-j} \to e$ as $j \to +\infty\}$; every horospherical subgroup is unipotent, and every maximal unipotent subgroup of a reductive Lie group is horospherical). For the case $G = SL_2(\mathbb{R})$, Conjectures 1 and 2 are provedn in [5] (for $\Gamma = SL_2(\mathbb{R})$) and in [8] (for arbitrary Γ); the main difference between the case $G = SL_2(\mathbb{R})$ and the general case consists in the fact that every connected unipotent subgroup of $SL_2(\mathbb{R})$ is horospherical. Also proven in [5] and [8] is the following:

3.4. THEOREM. *Let $G = SL_2(\mathbb{R})$, let Γ be a lattice in G and $x \in G/\Gamma$. Put $u_t = \begin{pmatrix} 1 & t \\ 0 & 1 \end{pmatrix}$ and $u = u_1 = \begin{pmatrix} 1 & 1 \\ 0 & 1 \end{pmatrix}$. Suppose that the orbit $\{u_t x | t \in \mathbb{R}\}$ is not periodic, i.e. $u_t x \neq x$ for every $t \neq 0$. Then the following hold:*

(1) *the orbit $\{u_t x | t \in \mathbb{R}\}$ is uniformly distributed with respect to the measure μ_G (defined in §2.5), i.e.*

$$\frac{1}{T} \int_0^T f(u_t x) \, dt \to \int_{G/\Gamma} f \, d\mu_G \quad \text{as} \quad T \to \infty,$$

for every bounded continuous function f on G/Γ;

(2) *the sequence $\{u^n x | n \in \mathbb{Z}\}$ is uniformly distributed with respect to μ_G, i.e.*

$$\frac{1}{N} \sum_{n=0}^{N-1} f(u^n x) \to \int_{G/\Gamma} f \, d\mu_G \quad \text{as} \quad N \to \infty,$$

for every bounded continuous function f on G/Γ.

Note that in (1) and (2), instead of bounded continuous functions, we could take characteristic functions of open sets whose boundary has measure zero.

3.5. Let φ be a homeomorphism of the locally compact metric space X. Then (1) a point $x \in X$ is called *recurrent* if there exists a sequence $\{n_k\}$ such that $\varphi^{n_k} x \to x$; (2) a point $x \in X$ is called *typical* if there exists a finite Borel measure μ_X on X such that

$$\lim_{n\to\infty} \frac{1}{n} \sum_{i=0}^{n-1} f(\varphi^i x) = \int_X f \, d\mu_x$$

for every bounded continuous function f on X. It follows from Birkhoff's ergodic theorem that if μ is a φ-invariant finite Borel measure on X, then μ-almost every point $x \in X$ is typical. If furthermore μ is strictly positive, then almost every point $x \in X$ is recurrent.

The following conjecture is formulated in [5] (for the case $G = SL_n(\mathbb{R})$ and $\Gamma = SL_n(\mathbb{Z})$).

CONJECTURE 3. Let G be a connected Lie group, Γ a lattice in G, and let $u \in G$. Suppose that u is unipotent, i.e. the transformation $\mathrm{Ad}\, u$ is unipotent. Then every point $x_0 \in G/\Gamma$ is recurrent and typical with respect to the homeomorphism $x \mapsto ux$ of G/Γ, where $x \in G/\Gamma$.

A proof of Conjecture 3 follows from Theorem 3.4 when $G = SL_2(\mathbb{R})$. It should be noted that Conjectures 1 and 3 are closely connected. Namely, if Conjecture 1 could be proved, then most likely Conjecture 3 could also be proved (and vice versa).

3.6. Suppose given a continuous action of a locally compact group on a locally compact space X. We say that a subgroup H of G has *property* (D) *with respect to* X if for every H-invariant locally finite Borel measure μ on X there exist Borel subsets X_i, $i \in \mathbb{N}^+$, such that 1) $\mu(X_i) < \infty$ for all i; 2) $\mu(X_i \triangle (hX_i)) = 0$ for all $h \in H$ and $i \in \mathbb{N}^+$; 3) $X = \bigcup_{i \in \mathbb{N}^+} X_i$. If H has property (D) with respect to X, then, as is easy to see, every H-ergodic, H-invariant, locally finite Borel measure on X is finite (when G and X are separable and metric, the converse is also true). In connection with Conjecture 1, the following theorem must be mentioned:

3.7. THEOREM (see [6]). *Let G be a connected Lie group and Γ a lattice in G. Then every unipotent subgroup U of G has property* (D) *with respect to G/Γ.*

3.8. LEMMA. *Suppose given a continuous action of a locally compact group G on a locally compact space X, and let H and F be subgroups of G. Suppose that the triple (G, H, F) has the Mautner property. Then the following hold*:

(1) *if μ is a Borel measure on X, $A \subset X$ is a Borel subset, $\mu(A) < \infty$ and $\mu(A \triangle (fA)) = 0$ for every $f \in F$, then $\mu(A \triangle (hA)) = 0$ for every $h \in H$*;

(2) *if F has property* (D) *with respect to X, then so does H;*

(3) *if $F \subset H$ and F has property* (D) *with respect to X, then every H-ergodic, H-invariant, locally finite Borel measure on X is F-ergodic.*

(1) follows from the equivalence of the conditions "$\rho(g)\chi_A = \chi_A$" and "$\mu(A \triangle (gA)) = 0$", where ρ is the unitary representation of G on $L_2(G, \mu_G)$ defined in §1.3, and χ_A is the characteristic function of the set A. (2) and (3) follow immediately from (1) and the definitions of property (D) and the Mautner property.

3.9. If H is a connected semisimple Lie group without compact factors and U is a maximal unipotent subgroup of H, then, as follows from Theorem 2.9, the triple (G, H, U) has the Mautner property, where G is any Lie group containing H. Therefore, Theorem 3.7 and Lemma 3.8 (2) imply the following:

3.10. THEOREM. *Let G be a connected Lie group, Γ a lattice in G, and let $H \subset G$ be a connected semisimple Lie group without compact factors. Then H has property* (D) *with respect to G/Γ.*

A special case of Theorem 3.10 is:

3.11. THEOREM. *Let G, Γ and H be as in Theorem 3.10 and let $x \in G/\Gamma$. Suppose that the orbit Hx is closed in G/Γ. Then $H \cap G_x$ is a lattice in H, where $G_x = \{g \in G | gx = x\}$ is the stabilizer subgroup of x.*

3.12. REMARKS. (1) In Theorems 3.10 and 3.11, the condition "H is a connected semisimple Lie group without compact factors" may be replaced by the condition "H is a connected Lie group for which any quotient of a radical by the nilpotent radical is compact."

(2) From Theorem 3.10 (using Remark (1) and setting $G = SL_n(\mathbb{R})$, $\Gamma = SL_n(\mathbb{Z})$), it is not difficult to derive the theorem of Borel and Harish-Chandra on the finiteness of the volume of quotient spaces of Lie groups by arithmetic subgroups.

(3) Using Theorem 2.9 and Lemma 3.8 (3), it is not difficult to show that if the condition "U is unipotent" in Conjecture 1 is replaced by the condition "U is generated by unipotent elements," then we obtain an equivalent conjecture. Note that if a quotient of some connected Lie group H by its nilpotent radical is semisimple and does not have compact factors, then H is generated by unipotent elements.

3.13. In view of Remark (3) of §3.12, as well as other considerations, it appears reasonable to generalize Conjecture 2 in the following way.

CONJECTURE $2'$. *Let G be a connected Lie group, Γ a lattice in G and H a subgroup of G. Suppose that H is generated by unipotent elements. Then* for every point $x \in G/\Gamma$ there exists a closed subgroup $P \subset G$ containing H such that the closure of the orbit Hx coincides with Px.

4. Applications to number theory and concluding remarks

4.1. If $t \in \mathbb{R}$, let $[t]$ be the largest integer not exceeding t, and let $\{t\} = t - [t]$. We denote by (m, n) the greatest common divisor of m and n, where m and n are positive integers. The following theorem is proved in [5], with the help of Theorem 3.4 (2).

4.2. THEOREM. *For every irrational number θ, we have*

$$\lim_{T \to \infty} \frac{1}{T} \sum_{\substack{0 < m \leq T\{m\theta\} \\ (m, [m\theta]) = 1}} \{m\theta\}^{-1} = \frac{1}{\zeta(2)} = \frac{6}{\pi^2},$$

where ζ denotes the Riemann zeta function.

4.3. Let B be a real, non-degenerate, indefinite quadratic form in n variables. It is well known that if $n \geq 5$ and the coefficients of B are rational, then B has a nontrivial zero, i.e. there are integers $x_1, \ldots, x_n$, not all zero, such that $B(x_1, \ldots, x_n) = 0$. Theorem 4.4 formulated below may be considered as an analog of this result for the case where B is not a multiple of a form with rational coefficients. Note that in Theorem 4.4 the condition "$n \geq 5$" is replaced by the condition "$n \geq 3$".

4.4. THEOREM (see [15]). *Suppose that $n \geq 3$ and B is not a multiple of a form with rational coefficients. Then given $\varepsilon > 0$, there exist integers $x_1, \ldots, x_n$, not all zero, such that $|B(x_1, \ldots, x_n)| < \varepsilon$.*

4.5. It follows easily from Theorem 4.4 that, under the same conditions, the set of values taken by B at integral points is dense in the set of real numbers.

It is easy to see that if Theorem 4.4 holds for some n_0, then it holds for all $n \geq n_0$. Hence it is enough to prove the theorem for $n = 3$. Note that if $n = 2$, then the analogous statement is false, as the example given by the form $x_1^2 - \lambda x_2^2$ shows, where λ is an irrational positive number such that the elements of the continued fractions expansion of $\sqrt{\lambda}$ are bounded.

Theorem 4.4 provides an answer to Davenport's conjecture (see [9]). This conjecture had previously been proved in the following cases: (a) $n \geq 21$ (see [10]); (b) $n = 5$ and B is of the form $B(x_1, \ldots, x_5) = \lambda_1 x_1^2 + \cdots + \lambda_5 x_5^2$ (see [10]). The proofs in [9] and [10] are based on methods of analytic number theory.

In [15], Theorem 4.4 is derived from Theorem 4.6 formulated below. Theorem 4.6 also gives an answer to a special case of Conjecture 2 of §3.10. A proof of this theorem, based on methods from the theory of algebraic groups and ergodic theory (more precisely, the theory of topological dynamical systems), is given in [15].

4.6. Theorem. *Let $G = SL_3(\mathbb{R})$, $\Gamma = SL_3(\mathbb{Z})$, and let H denote a group of elements of G that preserve the form $2x_1x_3 - x_2^2$. We denote by $G_x = \{g \in G | gx = x\}$ the stabilizer subgroup of x, where $x \in G/\Gamma$. If $x \in G/\Gamma$ and the orbit Hx is relatively compact in G/Γ, then the quotient space $H/H \cap G_x$ is compact.*

4.7. We show how Theorem 4.4 follows from Theorem 4.6. Denote by H_B the subgroup of G consisting of those matrices that preserve B. As remarked in §4.5, it is enough to prove Theorem 4.4 for $n = 3$. In this case $H = g_B H_B g_B^{-1}$ for some $g_B \in G$. Suppose that the conclusion of Theorem 4.4 is false. Then, by using Mahler's criterion for compactness, it is easy to show that the set $H_B \mathbb{Z}^3$ is relatively compact in G/Γ (we identify G/Γ with the space of lattices in $\mathbb{R}^3$). Using Theorem 4.6 with $x = g_B \mathbb{Z}^3$, we now obtain that the quotient space $H/H \cap G_x$ and, therefore, the quotient space $H_B/H_B \cap \Gamma$, is compact. Then, in view of Borel's density theorem (see [1]), the subgroup $H_B \cap \Gamma$ is Zariski-dense in H_B. Sub $\Gamma = SL_3(\mathbb{Z})$. Hence H_B is a $\mathbb{Q}$-subgroup of G, and therefore, B is a multiple of a form with rational coefficients. Contradiction.

By using methods of [15], Ragunathan's conjecture (Conjecture 2 of §3.3) can be proved when $G = SL_3(\mathbb{R})$ and the orbit Ux is relatively compact in G/Γ. This allows us to prove the following theorem, which can be considered a strengthening of Theorem 4.4.

4.8. Theorem. *Let B_1 and B_2 be real quadratic forms in 3 variables. Suppose that 1) for some basis of $\mathbb{R}^3$, B_1 and B_2 have the form $2x_1x_3 - x_2^2$ and x_1^2, respectively; 2) no non-trivial linear combination of B_1 and B_2 is a multiple of a form with rational coefficients. Then, given $\varepsilon > 0$, there are integers x_1, x_2, and x_3, not all zero, such that*

$$|B_1(x_1, x_2, x_3)| < \varepsilon \quad \text{and} \quad |B_2(x_1, x_2, x_3)| < \varepsilon.$$

4.9. Concluding remarks. In connection with the results of §2, we remark that an extensive survey of the ergodic properties of actions on homogeneous spaces is given in [3]. For the decomposition of such flows into ergodic components, see [19].

We did not touch upon many questions concerning connections between the theory of Lie groups and ergodic theory. In particular, we did not present results on the rigidity of discrete subgroups and ergodic actions (see [14], [21], [22]), on the connections between the finiteness of quotient groups of discrete subgroups and the invariant algebra of measurable subsets (see [12], [13]), and on the rigidity of horocycle flows (see [17], [18]).

Bibliography

1. A. Borel, *Density properties for certain subgroups of semisimple groups without compact components*, Ann. of Math. **72** (1960), 179–188.

2. P. Billingsley, *Ergodic theory and information*, Wiley, New York, 1965.
3. J. Brezin and C. C. Moore, *Flows on homogeneous spaces: a new look*, Amer. J. Math. **103** (1981), 571–613.
4. S. G. Dani, *Invariant measures and minimal sets of horospherical flows*, Invent. Math. **64** (1981), 357–385.
5. ____, *On uniformly distributed orbits of certain horocycle flows*, Ergodic Theory Dynamical Systems **2** (1982), 139–158.
6. ____, *On orbits of unipotent flows on homogeneous spaces*, Ergodic Theory and Dynamical Systems **4** (1984), 25–34.
7. ____, *Orbits of horospherical flows*, Duke Math. J. **53** (1986), 177–188.
8. S. G. Dani and J. Smillie, *Uniform distribution of horocycle orbits for Fuchsian groups*, Duke Math. J. **51** (1984), 185–194.
9. H. Davenport and H. Heilbronn, *On indefinite quadratic forms in five variables*, J. London Math. Soc. **21** (1946), 183–193.
10. H. Davenport and H. Ridout, *Indefinite quadratic forms*, Proc. London Math. Soc. **9** (1959), 544–555.
11. I. P. Kornfel'd, Ya. G. Sinai, and S. V. Fomin, *Ergodic theory*, "Nauka", Moscow, 1980.
12. G. A. Margulis, *Quotient groups of discrete subgroups and measure theory*, Funkstional. Anal. i Prilozhen. **12** (1978), no. 4, 64–75; English transl. in Functional Anal. Appl. **12** (1978).
13. ____, *Finiteness of quotient groups of discrete subgroups*, Funkstional. Anal. i Prilozhen. **13** (1979), no. 3, 28–39; English transl. in Functional Anal. Appl. **13** (1979).
14. ____, *Arithmeticity of irreducible lattices in semisimple groups of rank greater than one*, Appendix to the Russian transl. (Mir, Moscow, 1977) of M. C. Raghunathan, *Discrete subgroups of Lie groups*, Springer-Verlag, 1972.
15. G. A. Margulis, *Formes quadratiques indéfinies et flots unipotents sur les espaces homogènes*, C. R. Acad. Sci, 1987, 304 Sér. 1. n. 10. 294–253.
16. C. C. Moore, *The Mautner phenomenon for general unitary representations*, Pacif. J. Math. **86** (1980), 155–169.
17. M. Ratner, *Rigidity of horocycle flows*, Ann. of Math. **115** (1982), 597–614.
18. ____, *Horocycle flows, joinings and rigidity of products*, Ann. of Math. **118** (1983), 277–313.
19. A. N. Starkov, *The ergodic behavior of flows on homogeneous spaces*, Dokl. Akad. Nauk SSSR **273** (1983), no. 3, 538–540; English transl. in Soviet Math. Dokl. **28** (1983), 675–676.
20. A. A. Tempelman, *Ergodic theorems on groups*, Mokslas, Vil'nius 1986.
21. R. J. Zimmer, *Strong rigidity for ergodic actions of semisimple Lie groups*, Ann. of Math. **112** (1980), 511–529.
22. ____, *Ergodic theory and semisimple groups*, Birkhäuser, Boston, 1984.

Translated by V. DE ANGELIS

Lie Superalgebras of Vector Fields

A. L. ONISHCHIK

In recent years $\mathbb{Z}_2$-graded generalizations of classical algebra and analysis, which are sometimes called, with a certain measure of irony, "supermathematics", have become popular. Two stages can be recorded in the history of this mathematical direction. The first stage—the 40's and 50's—was inspired by the rapid development of algebraic topology. Here numerous examples of graded algebras appeared: cohomology algebras of manifolds, Pontryagin algebras of H-spaces, and others. This development relied significantly on the calculus of differential forms, which was developed considerably earlier by Poincaré, E. Cartan, and de Rham. The second stage began in the 70's and was inspired by the development of theoretical physics, which required an adequate mathematical apparatus. The concepts of superalgebra and supermanifold constituted the foundation of this apparatus (its development continues at present). Here the works of F. A. Berezin (see [1], [2]) have played a significant role.

The goal of the present lecture is to propose an introduction to a circle of ideas about which discussion has just begun. I shall attempt to accomplish this in the following example. One of the natural mathematical problems is the problem of finding the complete automorphism group of a given mathematical object, for example, the group of all symmetries of some ornament, the group of all motions of a given Riemannian space, the group of all automorphisms (that is, biholomorphic transformations) of a given complex analytic manifold, and others. We point out the following classical theorem: the automorphism group $\operatorname{Aut} \mathbb{C}P^n$ of the complex projective space $\mathbb{C}P^n$ coincides with the projective group $PGL_{n+1}(\mathbb{C})$. The group $\operatorname{Aut} M$ has also been found for the immediate generalizations of the manifold $\mathbb{C}P^n$—the so-called flag homogeneous spaces of M (see [6] and also [10]). In place of

---

1980 *Mathematics Subject Classification* (1985 *Revision*). Primary 17A70, 17B10, 17B70; Secondary 58C99, 83E99, 58A05, 32L10.

the automorphism group $\operatorname{Aut} M$ of a compact complex manifold M, one can examine its Lie algebra consisting of all holomorphic vector fields on M (with some loss of information, since the tangent algebra determines only the connected component of the identity of a Lie group). On the other hand, with each compact complex supermanifold one can associate a finite-dimensional Lie superalgebra of "vector fields" on this supermanifold, which can naturally be interpreted as "infinitesimal automorphisms" of it. The topic of discussion will be the calculation of Lie superalgebras of vector fields for certain supermanifolds constructed in flag homogeneous spaces.

All vector spaces and algebras will be assumed to be defined over the field $\mathbb{C}$, and all manifolds will be complex analytic. Of course, many definitions carry over trivially to other fields or to real differentiable manifolds.

1. Algebraic concepts

Below we shall consider vector spaces and algebras graded by means of the group $\mathbb{Z}_2 = \{\bar{0}, \bar{1}\}$. If

$$V = V_{\bar{0}} \oplus V_{\bar{1}}$$

is a $\mathbb{Z}_2$-graded space (or superspace), then the elements of $V_{\bar{0}}$ are called even, and the elements of $V_{\bar{1}}$ are called odd. The parity function $p(x)$ is defined on homogeneous elements by the formula

$$p(x) = k \quad \text{if} \quad x \in V_{\bar{k}}, \quad k = 0, 1.$$

Any $\mathbb{Z}$-graded space

$$V = \bigoplus_{k \in \mathbb{Z}} V_k$$

can be converted to a $\mathbb{Z}_2$-graded space by "reduction modulo 2":

$$V = V_{\bar{0}} \oplus V_{\bar{1}}, \quad \text{where} \quad V_{\bar{0}} = \bigoplus_{k \in \mathbb{Z}} V_{2k}, \quad V_{\bar{1}} = \bigoplus_{k \in \mathbb{Z}} V_{2k+1}.$$

We denote by $\mathbb{C}^{p|q}$ the standard superspace $\mathbb{C}^p \oplus \mathbb{C}^q$, where the left summand consists of even elements and the right of odd elements.

A $\mathbb{Z}_2$-graded algebra, that is, an algebra of the form

$$A = A_{\bar{0}} \oplus A_{\bar{1}},$$

where $A_p A_q \subset A_{p+q}$ ($p, q \in \mathbb{Z}_2$), is called a superalgebra. Any $\mathbb{Z}$-graded algebra becomes a superalgebra after a gradation reduction modulo 2. (We shall call superalgebras constructed in this way graded.)

A superalgebra A is called commutative if

$$ba = (-1)^{p(a)p(b)} ab$$

for any homogeneous $a, b \in A$.

EXAMPLES. 1. An exterior algebra (Grassmann algebra) $A = \Lambda(\xi_1, \ldots, \xi_n)$ with generators $\xi_1, \ldots, \xi_n$ is a commutative graded superalgebra with $\mathbb{Z}$-graded subspaces

$$A_p = \Lambda^p(\xi_1, \ldots, \xi_n) \quad (p \in \mathbb{Z}).$$

2. The algebra $L(V)$ of all linear transformations of a $\mathbb{Z}_2$-graded space V becomes a superalgebra if we set
$$L(V_k) = \{X \in L(V) | X(V_l) \subset V_{l+k} \ (l \in \mathbb{Z}_2)\}, \qquad k \in \mathbb{Z}_2.$$
If a basis $e_1, \ldots, e_p, f_1, \ldots, f_q$ is chosen in V so that $V_{\bar{0}} = \langle e_1, \ldots, e_p \rangle$ and $V_{\bar{1}} = \langle f_1, \ldots, f_q \rangle$, then linear transformations can be identified with matrices

$$p \quad q \quad pq \quad X = \left(\begin{array}{c|c} X_{00} & X_{01} \\ \hline X_{10} & X_{11} \end{array} \right) \begin{array}{c} p \\ q \end{array} \qquad (1)$$

where $p(X) = 0$ if $X_{01} = 0$, $X_{10} = 0$ and $p(X) = 1$ if $X_{00} = 0$, $X_{11} = 0$. We denote the matrix superalgebra by $L_{p|q}(\mathbb{C}) = L(\mathbb{C}^{p|q})$.

3. In the superalgebra $L(V)$ from Example 2 we replace multiplication by the commutator, which is defined on homogeneous elements by the formula
$$[X, Y] = XY - (-1)^{p(X)p(Y)} YX. \qquad (2)$$
We then obtain a superalgebra $gl(V)$ with the following properties:
$$[Y, X] = (-1)^{p(X)p(Y)+1}[X, Y],$$
$$[X, [Y, Z]] = [[X, Y], Z] + (-1)^{p(X)p(Y)}[Y[X, Z]]. \qquad (3)$$

Let $gl_{p|q}(\mathbb{C}) = gl(\mathbb{C}^{p|q})$ be the corresponding matrix superalgebra. We note that the superalgebras in Examples 2 and 3 are graded. For example, $gl_{p|q}(\mathbb{C}) = \mathfrak{g}_{-1} \oplus \mathfrak{g}_0 \oplus \mathfrak{g}_1$, where $\mathfrak{g}_0 = \mathfrak{g}_{\bar{0}}$ is defined by the conditions $X_{01} = 0$, $X_{10} = 0$, $\mathfrak{g}_{-1}$ by the conditions $X_{00} = 0$, $X_{10} = 0$, $X_{01} = 0$, and $\mathfrak{g}_1$ by the conditions $X_{00} = 0$, $X_{01} = 0$, $X_{11} = 0$ (see [3]).

A superalgebra is called a Lie superalgebra if the multiplication [,] in it satisfies conditions (3). Generalizing Example 3, one can convert any associative superalgebra to a Lie superalgebra by defining a new multiplication in it by means of (2). The construction expounded in the next example is extremely important.

EXAMPLE 4. Let A be a superalgebra. We define the $\mathbb{Z}_2$-graded subspace $\operatorname{Der} A \subset L(A)$ of derivations of A as follows:
$$\operatorname{Der}_k A = \{X \in L(A)_k | X_{(a,b)} = (X_a)b + (-1)^{p(X)p(a)} a(Xb) \text{ for } a, b \in A\}.$$

Then $\operatorname{Der} A$ is a subalgebra in $gl(A)$. Thus, with each superalgebra A is associated its Lie superalgebra of derivations $\operatorname{Der} A$. If A is a graded superalgebra, then $\operatorname{Der} A$ is also a graded superalgebra.

In particular, we set $W_n = \operatorname{Der} \Lambda(\xi_1, \ldots, \xi_n)$. This graded Lie superalgebra is unusual in that it is simple for $n \geq 2$, that is, contains no proper nonzero $\mathbb{Z}_2$-graded ideals [3].

2. Supermanifolds

The "super" generalization of classical analysis studies functions that take their values in some commutative superalgebra, for example, in a Grassmann algebra. Let $\varphi: U \to \Lambda(\xi_1, \ldots, \xi_n)$ be a holomorphic function in a domain $U \subset \mathbb{C}^p$. Then φ in a neighborhood of any point z_0 of U can be expanded into a power series in $x_1, \ldots, x_p, \xi_1, \ldots, \xi_q$, where the x_i are any local coordinates in a neighborhood of z_0 such that $z_0 = (0, \ldots, 0)$. Therefore, we can assume that φ is an analytic function in the "even coordinates" $x_1, \ldots, x_p$ and the "odd coordinates" $\xi_1, \ldots, \xi_q$ and write it in the form $\varphi = \varphi(x_1, \ldots, x_p, \xi_1, \ldots, \xi_q)$. Let $\mathscr{F}$ be the sheaf of germs of ordinary holomorphic functions in U, and let $\Lambda_{\mathscr{F}}(\xi_1, \ldots, \xi_q) = \mathscr{F} \otimes \Lambda(\xi_1, \ldots, \xi_q)$ be the sheaf of holomorphic functions with values in $\Lambda(\xi_1, \ldots, \xi_q)$.

The term "supermanifold" will be used in the sense of Berezin-Leĭtes (see [2], [4]), but here, as in [5], the complex analytic case will be examined. A twistor space $(M, \mathscr{O})$ over the field $\mathbb{C}$ locally isomorphic to a space of the form $(U, \Lambda_{\mathscr{F}}(\xi_1, \ldots, \xi_q))$, where U is an open set in $\mathbb{C}^p$, is called a supermanifold (more precisely, a complex supermanifold). The sheaf $\mathscr{O}$ is called a structure sheaf. The morphisms and automorphisms of supermanifolds are understood in the sense of the theory of twistor spaces (see [5]).

Just as an ordinary manifold, a supermanifold can be defined by an atlas in the topological space M. In each chart of this atlas the "even coordinates" $x_1, \ldots, x_p$ and "odd coordinates" $\xi_1, \ldots, \xi_q$ are defined; in the intersection with another chart where the coordinates $\tilde{x}_i, \tilde{\xi}_j$ are defined, expressions of the form

$$\tilde{x}_i = \varphi_i(x_1, \ldots, x_p, \xi_1, \ldots, \xi_q),$$
$$\tilde{\xi}_j = \psi_j(x_1, \ldots, x_p, \xi_1, \ldots, \xi_q)$$

must hold, where φ_i and ψ_j are analytic functions (transitions functions).

EXAMPLE 1. This example has actually been examined already in the works of E. Cartan. Let M be an arbitrary complex manifold of dimension n, and let $\mathscr{O} = \Omega$ be the sheaf of germs of holomorphic exterior forms on M. Then locally $\Omega = \Lambda_{\mathscr{F}}(dx_1, \ldots, dx_n)$, where $x_1, \ldots, x_n$ are local coordinates and $\mathscr{F}$ is the sheaf of germs of holomorphic functions on M. We therefore obtain a supermanifold of dimension $n|n$ with "even" local coordinates x_i and "odd" local coordinates $\xi_j = dx_j$. If $\tilde{x}_1, \ldots, \tilde{x}_n$ is another local coordinate system on M and $\tilde{\xi}_j = d\tilde{x}_j$, then the transition functions have the form

$$\tilde{x}_i = \varphi_i(x_1, \ldots, x_n), \qquad \tilde{\xi}_j = \sum_k \frac{\partial \varphi_j}{\partial x_k} \xi_k.$$

Thus, the φ_i do not depend on ξ_j, and the ψ_j are linear in ξ_k.

The special form of the transition functions in Example 1 is related to the fact that the sheaf $\mathscr{O}$ has the following global structure there: $\mathscr{O} = \Lambda_{\mathscr{F}} \Omega^1$, where Ω^1 is the sheaf of germs of Pfaffian forms, that is, cross-sections of the tangent bundle. In general, let E be a holomorphic vector bundle over a complex manifold M, and let $\mathscr{E}$ be the sheaf of germs of its holomorphic cross-sections. Setting $\mathscr{O} = \Lambda_{\mathscr{F}} \mathscr{E}$, we obtain a supermanifold $(M, \mathscr{O})$.

Supermanifolds of this kind (and those isomorphic to them) are called split. We note that a structure sheaf $\mathcal{O}$ of a split supermanifold is $\mathbb{Z}$-graded:

$$\mathcal{O} = \bigoplus_{k \in \mathbb{Z}} \mathcal{O}_k, \qquad \text{where} \quad \mathcal{O}_k = \Lambda^k_{\mathcal{F}}\mathcal{E}. \tag{4}$$

Any supermanifold of dimension $p|1$ is split. Any differentiable (real) supermanifold is split as well.

EXAMPLE 2. The following important class of split supermanifolds is connected with homogeneous spaces. Let G be a complex Lie group, P a Lie subgroup of G, $M = G/P$, and $\rho: P \to GL(V)$ a finite holomorphic linear representation. Then there arises a holomorphic vector bundle E_ρ with bases M and fiber V associated with the fibering of G into cosets relative to P. We obtain a split supermanifold $(M, \Lambda_{\mathcal{F}}\mathcal{E}_\rho)$.

EXAMPLE 3. We present the example of a nonsplit supermanifold (super-Grassmannian) constructed by Yu. I. Manin [5]. It is convenient to construct it not with the help of a structure sheaf, but by gluing together local models. Let us first recall the construction of an ordinary Grassmannian in the example of the manifold $G_{4,2}$. This is the manifold of all two-dimensional subspaces in $\mathbb{C}^4$. Any point $x \in G_{4,2}$ can be defined by a matrix

$$X = \begin{pmatrix} x_{11} & x_{12} \\ x_{21} & x_{22} \\ x_{31} & x_{32} \\ x_{41} & x_{42} \end{pmatrix} \tag{5}$$

of rank 2. We call the numbers $x_{ij} \in \mathbb{C}$ homogeneous coordinates of x. They are not defined uniquely. More precisely, two matrices X and $\tilde{X}$ define the same point x if and only if $\tilde{X} = XA$, where $A \in GL_2(\mathbb{C})$. We now define an atlas on $G_{4,2}$ as follows. For each pair (i, j) $(1 \leq i < j \leq 4)$ we set

$$U_{ij} = \left\{ x \in G_{4,2} \left\| \begin{matrix} x_{i1} & x_{i2} \\ x_{j1} & x_{j2} \end{matrix} \right\| \neq 0 \right\}.$$

If $x \in U_{ij}$, then in the matrix $Y = X \bigl(\begin{smallmatrix} x_{i1} & x_{i2} \\ x_{j1} & x_{j2} \end{smallmatrix} \bigr)^{-1}$ the ith and jth rows contain the unit matrix, and the remaining 4 elements are uniquely defined and are taken for the local (nonhomogeneous) coordinates of x.

The super-Grassmannian $G_{n|m,k|l}$ $(0 \leq m \leq n, 0 \leq l \leq k)$, whose reduction (see below) coincides with $G_{n,k} \times G_{m,l}$, is constructed in a completely analogous manner. We consider the case $n = m = 2$ and $k = l = 1$. A point of the manifold $G_{2,1} \times G_{2,1}$ represents a graded subspace of dimension $1|1$ in $\mathbb{C}^{2|2}$. Therefore, it is defined by a matrix X of the form (5), where $x_{12} = x_{22} = x_{31} = x_{41} = 0$. Let us imagine that we have extended the scalar domain $\mathbb{C}$ of the space $\mathbb{C}^{2|2}$ to some commutative superalgebra over $\mathbb{C}$. Then an even vector will have some odd coordinates for the odd elements of the basis, and an odd vector will have some even coordinates for the even elements of the basis. Therefore, we shall define a "point" x of the super-Grassmannian by a matrix X over the ring $A = \mathbb{C}(x_{11}, x_{12}, x_{21}, x_{22}, \xi_{11}, \xi_{12}, \xi_{21}, \xi_{22})$ of the form

$$X = \begin{pmatrix} x_{11} & \xi_{12} \\ x_{21} & \xi_{22} \\ \hline \xi_{11} & x_{12} \\ \xi_{21} & x_{22} \end{pmatrix}. \tag{6}$$

Here it is assumed that at least one of the even submatrices of order 2 is invertible. This is equivalent to the fact that one of the following four conditions holds:

$$x_{11}x_{12} \neq 0, \quad x_{21}x_{12} \neq 0, \quad x_{11}x_{22} \neq 0, \quad x_{21}x_{22} \neq 0. \tag{7}$$

We shall assume that a "point" x does not change under multiplication of the matrix X on the right by a matrix of the form $\begin{pmatrix} u_1 & v_1 \\ v_2 & u_2 \end{pmatrix}$, where $u_1, u_2 \in A_{\bar{0}}$, $v_1, v_2 \in A_{\bar{1}}$, and u_1, u_2 are invertible in A. With each of the conditions (7) we associate a chart on the super-Grassmannian, operating exactly as in the classical case. For example, if $x_{21}x_{12} \neq 0$, then we obtain

$$X \begin{pmatrix} x_{21} & \xi_{22} \\ \xi_{11} & x_{12} \end{pmatrix}^{-1} = \begin{pmatrix} x & \xi \\ 1 & 0 \\ \hline 0 & 1 \\ \eta & y \end{pmatrix}.$$

In the corresponding chart x, y are assumed to be even coordinates, and ξ, η odd coordinates.

In [5] it is proved that the supermanifold $G_{2|2,1|1}$ is nonsplit. At the same time the supermanifolds $G_{n|m,k|0}$ and $G_{n|m,0|l}$ (and also $G_{n|m,k|m}$ and $G_{n|m,n|l}$) are split and can be obtained by means of the construction of Example 2 applied to $M = G_{n,k}$ or $G_{m,l}$.

For an arbitrary supermanifold $(M, \mathcal{O})$ of dimension $p|q$ the structure sheaf $\mathcal{O}$ admits a filtration

$$\mathcal{O} \supset \mathcal{J} \supset \mathcal{J}^2 \supset \cdots \supset \mathcal{J}^q \supset \mathcal{J}^{q+1} = 0, \tag{8}$$

where $\mathcal{J}$ is the ideal in $\mathcal{O}$ generated by odd elements. We consider the corresponding graded sheaf

$$\operatorname{gr} \mathcal{O} = \bigoplus_{s=0}^{q} \operatorname{gr}_s \mathcal{O}, \qquad \text{where} \quad \operatorname{gr}_s \mathcal{O} = \mathcal{J}^s / \mathcal{J}^{s+1}.$$

The sheaf $\mathcal{F} = \operatorname{gr}_0 \mathcal{O}$ defines on M the structure of a complex analytic manifold M_{rd} (reduction of the supermanifold $(M, \mathcal{O})$), $\mathcal{E} = \operatorname{gr}_1 \mathcal{O}$ is a locally free analytic sheaf on M_{rd}, and $\operatorname{gr} \mathcal{O} = \Lambda_{\mathcal{F}} \mathcal{E}$. Thus, $(M_{\mathrm{rd}}, \operatorname{gr} \mathcal{O})$ is a split supermanifold (see [5]).

3. Vector fields on supermanifolds

Let $(M, \mathcal{O})$ be a supermanifold. We denote by $\operatorname{Der} \mathcal{O}$ the sheaf of derivations of $\mathcal{O}$. This is a sheaf of Lie superalgebras; the filtration (7) generates a filtration in $\mathrm{Der}\mathscr{O}$. The Lie superalgebra $\theta = \Gamma(M, \mathrm{Der}\mathscr{O})$ of sections of the sheaf $\mathrm{Der}\mathscr{O}$ is called the superalgebra of vector fields on $(M, \mathscr{O})$. If M is compact, then $\dim \theta < \infty$.

If the supermanifold $(M, \mathscr{O})$ is split, then the sheaf $\mathrm{Der}\mathscr{O}$ is the sheaf of graded Lie superalgebras, and θ is a graded Lie superalgebra:

$$\theta = \bigoplus_{S=-1}^{q} \theta_S.$$

EXAMPLE 1. We consider the supermanifold (M, Ω) from Example 1 in §2. Let $\mathfrak{g}$ be the Lie algebra of holomorphic vector fields on M. Each vector field $v \in \mathfrak{g}$ defines a derivation of degree 0 of the sheaf Ω—the so-called Lie derivative—which we shall also denote by v. Thus, $\mathfrak{g}$ can be embedded into θ_0 as a superalgebra. On the other hand, a field $v \in \mathfrak{g}$ defines a derivation $\hat{v} \in \theta_{-1}$—inner multiplication on v. These derivations exhaust θ_{-1}. In θ_0 there is also a grading derivation ε defined by the formula $\varepsilon(\varphi) = s\varphi$ for $\varphi \in \mathscr{O}_s$. Finally, the exterior differential d is an element of the space θ_1. As is known, the following relations hold:

$$[v, \hat{w}] = [v, w], \qquad [\hat{v}, \hat{w}] = [d, v] = 0,$$
$$[d, \hat{v}] = v, \qquad [d, d] = 0.$$

It follows that $\tilde{\theta} = \theta_{-1} \oplus (\mathfrak{g} \oplus \langle \varepsilon \rangle) \oplus \langle d \rangle$ is a superalgebra of θ. The following theorem is valid:

THEOREM 1 [7]. *If M is the flag homogeneous space of a simple complex Lie group, then the Lie superalgebra of vector fields on (M, Ω) coincides with $\tilde{\theta}$.*

EXAMPLE 2. We consider the super-Grassmannian $G_{n|m, k|l}$ from Example 3 in §2. Then there is a natural Lie superalgebra homomorphism $\Theta: gl_{n|m}(\mathbb{C}) \to \theta$ whose kernel is the ideal $\langle E \rangle$ consisting of scalar matrices. This homomorphism is defined by the action of the Lie supergroup $GL_{n|m}$ on $G_{n|m, k|l}$ by left multiplication of matrices of the form (6) by matrices in $GL_{n|m}$. (Concerning Lie supergroups and their actions see [4].) The homomorphism Θ makes it possible to identify the Lie superalgebra $pgl_{n|m}(\mathbb{C}) = gl_{n|m}(\mathbb{C})/\langle E \rangle$ with a subalgebra in θ.

THEOREM 2. *The Lie superalgebra θ coincides with $pgl_{n|m}(\mathbb{C})$ in all cases except $m = k = l = 1$; $m = 1$, $k = n-1$, $l = 0$; $k = 1$, $l = m-1 \geq 1$; $k = n-1 \geq 1$, $l = 1$.*

This result is due to A. A. Serov. The proof essentially uses the passage to the split supermanifold $(M_{\mathrm{rd}}, \mathrm{gr}\mathscr{O})$.

EXAMPLE 3. We consider the split supermanifold $G_{n|1, n-1|0} (\simeq G_{n|1, 1|1})$, which is an exception in Theorem 2.

THEOREM 3. *The Lie superalgebra θ of vector fields on $G_{n|1,n-1|0}$ is isomorphic to W_n (see Example* 4 *in* §1).

Let us describe the action of the Lie superalgebra W_n on $G_{n|1,n-1|0}$ discussed in this theorem. It is not hard to show that $G_{n|1,n-1|0}$ is split and isomorphic to the supermanifold $(G_{n,n-1}, \Lambda_{\mathscr{F}}\mathscr{E})$, where E is a tautological vector bundle over $G_{n,n-1}$ whose fiber over a point $x \in G_{n,n-1}$ is the point x itself, regarded as an $(n-1)$-dimensional vector space. It will be convenient for us to interpret $G_{n,n-1}$ as the $(n-1)$-dimensional projective space $P(\mathbb{C}^{n*})$ associated with the space $\mathbb{C}^{n*}$ conjugate to $\mathbb{C}^n$. Moreover, a line $\langle u \rangle \subset \mathbb{C}^{n*}$, where $u \neq 0$, is identified with the $(n-1)$-dimensional subspace $\operatorname{Ker} u \subset \mathbb{C}^n$. Then E is the submanifold in $P(\mathbb{C}^{n*}) \times \mathbb{C}^n$ consisting of pairs $(\langle u \rangle, v)$ such that $u(v) = 0$. Let $e_1, \ldots, e_n$ be the standard basis in $\mathbb{C}^n$. We shall write an element $u \in \mathbb{C}^{n*}$ in the form $u = (u_1, \ldots, u_n)$, where $u_i = u(e_i)$. We shall assume that the elements $u_1, \ldots, u_n$ are even and call them homogeneous coordinates on our super-Grassmannian, and we shall assume that the elements $e_1, \ldots, e_n$ are odd. We consider the superalgebra

$$A = \Lambda(e_1, \ldots, e_n) \otimes \mathbb{C}[u_1, \ldots, u_n].$$

The elements $\xi_{ij} = u_i e_j - u_j e_i \in A$ $(i \neq j)$ can be called odd homogeneous coordinates on the super-Grassmannian. Over the open set $\{u_i \neq 0\} \subset P(\mathbb{C}^{n*})$ the elements $(1/u_i)\xi_{ij} = e_j - (u_j/u_i)e_i$ $(j \neq i)$, obviously, form a basis of holomorphic sections of the bundle E. We call the elements u_i/u_j and $(1/u_i)\xi_{ij}$ $(j \neq i)$ of the algebra

$$B = \Lambda(e_1, \ldots, e_n) \otimes \mathbb{C}(u_1, \ldots, u_n)$$

nonhomogeneous coordinates in the domain $\{u_i \neq 0\}$. These are, in fact, local coordinates of the super-Grassmannian in this domain. The structure sheaf $\mathscr{O} = \Lambda_{\mathscr{F}}\xi$ consists of local analytic functions in these coordinates. Let C be the subalgebra in B generated by u_j/u_i and $(1/u_i)\xi_{ij}$ for all $i \neq j$. It is easy to verify that $C = \operatorname{Ker}\Delta$, where $\Delta \in \operatorname{Der}_{\bar{1}} B$ is the derivation defined by the formulas

$$\Delta e_i = u_i, \qquad \Delta u_i = 0 \quad (i = 1, \ldots, n).$$

Let $\delta \in W_n = \operatorname{Der}\Lambda(e_1, \ldots, e_n)$. Then δ can be uniquely extended to a derivation of the superalgebra B, which is denoted by the same letter and satisfies the condition $[\delta, \Delta] = 0$ [3]. Obviously, $\delta(C) \subset C$. Therefore, δ defines a derivation $\tilde{\delta}$ of the sheaf $\mathscr{O}$. The correspondence $\delta \mapsto \tilde{\delta}$ is an injective homomorphism of the graded Lie superalgebra W_n into $\theta = \Gamma(P(\mathbb{C}^{n*}), \operatorname{Der}\mathscr{O})$, which, in fact, turns out to be surjective.

4. Transitive irreducible graded Lie superalgebras

When proving theorems similar to the theorems in §3, it is useful to know in advance in which class of Lie superalgebras one should seek the unknown superalgebra of vector fields. For example, in the similar problem of calculating the Lie algebra of holomorphic vector fields on a flag homogeneous space G/P, one can *a priori* prove that this algebra is simple if G is a simple Lie group. This makes it possible to use the well-known classification of simple Lie algebras (see [6]). Not all of the Lie superalgebras of vector fields that appear in Theorems 1, 2, and 3 are simple. However, they all belong to another well-known class of Lie superalgebras, which we shall now define.

A graded Lie superalgebra $\mathfrak{g} = \bigoplus_{s \geq -1} \mathfrak{g}_s$ is called transitive if for any $x \in \mathfrak{g}_s$, where $s \geq 0$, there exists $y \in \mathfrak{g}_{-1}$ such that $[x, y] \neq 0$, and irreducible if the adjoint representation of the Lie algebra $\mathfrak{g}_0$ and $\mathfrak{g}_{-1}$ is irreducible.

Finite-dimensional, complex, transitive, irreducible graded Lie superalgebras were classified by V. G. Kac [3] (see also [8]). We shall briefly dwell on this classification. If a finite-dimensional graded Lie superalgebra $\mathfrak{g}$ over $\mathbb{C}$ is transitive and irreducible, then the adjoint representation of $\mathfrak{g}_0$ in $\mathfrak{g}_{-1}$ is irreducible and faithful. Therefore, either $\mathfrak{g}_0$ is semisimple, or $\mathfrak{g}_0 = \mathfrak{g}_0' \oplus \langle \varepsilon \rangle$, where $\mathfrak{g}_0' = [\mathfrak{g}_0, \mathfrak{g}_0]$ and $\varepsilon \neq 0$ lies in the center of the algebra $\mathfrak{g}_0$. In the latter case $[\varepsilon, x] = sx$ for all $x \in \mathfrak{g}_s$. In what follows we shall assume that $\mathfrak{g}_0$ is not semisimple (in fact, this can always be achieved by joining the one-dimensional center to $\mathfrak{g}_0$). It turns out that finite-dimensional, transitive, irreducible, complex Lie superalgebras $\mathfrak{g}$ with semisimple $\mathfrak{g}_0$ can be divided into the following four types:

1. $\mathfrak{g} = \mathfrak{g}_{-1} \oplus \mathfrak{g}_0$, where $\mathfrak{g}_0$ acts in $\mathfrak{g}_{-1}$ faithfully and irreducibly.

2. $\mathfrak{g} = \mathfrak{g}_{-1} \oplus \mathfrak{g}_0 \oplus \mathfrak{g}_1$, where $\mathfrak{g}_0 = \mathfrak{a} \oplus \langle \varepsilon \rangle$ and $\mathfrak{g}_{-1} = \hat{\mathfrak{a}} \simeq \mathfrak{a}$; $\mathfrak{a}$ is a simple complex Lie algebra and $\mathfrak{g}_{-1} = \langle d \rangle$, where the commutator in $\mathfrak{g}$ is as in the superalgebra $\tilde{\theta}$ from Example 1 in §3.

3. The following Lie superalgebras, which are close to being simple:

a) $\mathfrak{g} = pgl_{n|m}$ $(n + m \geq 2)$;

b) $\mathfrak{g} = OSP_{2|2q} \subset gl_{2|2q}$ $(q \geq 1)$ is the superalgebra consisting of matrices of the form

$$\begin{pmatrix} \alpha & 0 & u_1 & u_2 \\ 0 & -\alpha & v_1 & v_2 \\ \hline v_2^T & u_2^T & X & Y \\ -v^T & -u_1^T & Z & -X^T \end{pmatrix} \begin{matrix} 1 \\ 1 \\ q \\ q \end{matrix}, \qquad Y^T = Y, \qquad Z^T = Z;$$

c) $\mathfrak{g} = \pi_n$, $\pi_n^* \subset gl_{n|n}$ $(n \geq 2)$ are the subalgebras consisting, respectively, of matrices of the form

$$\begin{pmatrix} X & Y \\ \hline Z & -X^T \end{pmatrix} \begin{matrix} n \\ n \end{matrix}, \quad Y^T = Y, \quad Z^T = -Z \quad \text{or} \quad Y^T = -Y, \quad Z^T = Z.$$

4. The following Lie superalgebras of Cartan type (see [3]):

a) $\mathfrak{g} = W_n$ $(n \geq 2)$;

b) $\mathfrak{g} = S_n \oplus \langle \varepsilon \rangle$ $(n \geq 2)$, where $S_n \subset W_n$ is the subalgebra of derivations with zero divergence;

c) $\mathfrak{g} = H_n \oplus \langle \varepsilon \rangle$ and $\widetilde{H}_n \oplus \langle \varepsilon \rangle$ $(n \geq 3)$, where $\widetilde{H}_n \subset W_n$ is the subalgebra of Hamiltonian derivatives and $H_n = [\widetilde{H}_n, \widetilde{H}_n]$.

It turns out that the following transitivity criterion for a superalgebra of vector fields holds:

THEOREM 4 [7]. *Let $(M, \Lambda_{\mathscr{F}}\mathscr{E})$ be a split supermanifold, where the bundle E^* is generated by global holomorphic sections, and let θ be the Lie superalgebra of vector fields on $(M, \Lambda_{\mathscr{F}}\mathscr{E})$. If the adjoint representation of the Lie algebra θ_0 in θ_{-1} is faithful, then θ is transitive.*

The following theorem can be deduced from this:

THEOREM 5 [7]. *Let G be a connected, semisimple, complex Lie group and P a parabolic subgroup of it, where the natural homomorphism $G \to \mathrm{Aut}^{\circ} G/P$ is surjective, and let $\rho\colon G \to GL(V)$ be an irreducible holomorphic representation of P. In G we choose a Borel subgroup B such that P contains the opposite Borel subgroup B_-, and we denote by λ the highest weight of the representation ρ^* with respect to B. If λ is a dominant weight of the group G that is not equal to 0 on any of its simple components, then the Lie superalgebra of vector fields on the supermanifold $(M, \Lambda_{\mathscr{F}}\mathscr{E}_\rho)$ (see Example 2 in §2) is transitive and irreducible. Moreover, $\theta_0 \simeq \mathfrak{g} \oplus \langle \varepsilon \rangle$, where $\mathfrak{g}$ is the tangent algebra of the group G and θ_{-1} is an irreducible $\mathfrak{g}$-module with highest weight λ.*

We note that in A. A. Serov's dissertation (see [9]) the explicit form of the Lie superalgebras θ is found in all the cases described in Theorem 5. Theorems 4 and 5 and the above classification of V. G. Kac are also used in the proof of Theorems 1, 2, and 3 in §3. An important role in the proof of all these results is played by the techniques of induced representations of the group G in the sections and cohomologies connected with homogeneous bundles on G/P.

BIBLIOGRAPHY

1. F. A. Berezin, *Introduction to algebra and analysis with anticommuting variables*, Moscow, 1983; rev. aug. English transl., *Introduction to superanalysis*, Reidel, Dordrecht, 1987.

2. F. A. Berezin and D. A Leites, *Supermanifolds*, Dokl. Akad. Nauk SSSR **224** (1975), 505–508; English transl. in Soviet Math. Dokl. **16** (1975), 1218–1222.

3. V. G. Kac, *Lie superalgebras*, Adv. in Math. **26** (1977), 8–96.

4. D. A. Leites, *Theory of supermanifolds*, Petrozavodsk, 1983. (Russian)

5. Yu. I. Manin, *Gauge fields and complex geometry*, "Nauka", Moscow, 1984; English transl., Springer-Verlag, Berlin and New York, 1988.

6. A. L. Onishchik, *Inclusion relations among transitive compact transformation groups*, Trudy Moskov. Mat. Obshch. **11** (1962), 199–242; English Transl. in Amer. Math. Soc. Transl. (2) **50** (1966), 5–58.

7. ___, *Transitive Lie superalgebras of vector fields*, Yarosl. Univ., Yaroslavl, 1986. 24 pp. Deposited in VINITI 06/12/86, no. 4329-V. (Russian)

8. M. Scheunert, *The theory of Lie superalgebras*, Berlin, 1979.

9. A. A. Serov, *Lie superalgebras of vector fields on complex flag supermanifolds*, Yarosl. Univ., Yaroslavl, 1986. 55 pp. Deposited in VINITI 01/26/87, no. 610-V. (Russian)

10. M. Steinsieck, *Transformation groups on homogeneous rational manifolds*, Math. Ann. **260** (1982), 423–435.

Translated by R. LENET

Invariant Theory

V. L. POPOV

Invariant theory, which already has an almost 150 year history, is now undergoing a period of new development: a satisfactory understanding of the general picture in the classical situation has been achieved and new directions, methods, and applications have been discovered. In essence, it is now identified with the theory of actions of algebraic groups on algebraic varieties (schemes), the local aspect of which pertaining to actions on affine varieties is most closely connected with invariant theory in the classical sense and so is the most developed.

In this report I will discuss mainly the principal results centered around the classical theme of describing invariants and covariants; consideration of other aspects and applications of modern invariant theory would require a separate report.

1. Classical problems

Suppose k is an algebraically closed field, V is a finite-dimensional vector k-space, $k[V]$ is the algebra of regular functions (polynomials) on V, and G is an algebraic subgroup of $GL(V)$. The group G acts naturally on $k[V]$.

The main problem of the classical theory is to "explicitly describe" the algebra of invariants $k[V]^G$. The idea of the description is as follows: a) see whether $k[V]^G$ has a finite system of generators; b) if it does, give a constructive method for finding a minimal system (ideally, find it explicitly).

The problem of describing those G for which $k[V]^G$ is finitely generated is known as the original Hilbert's 14th problem (Hilbert himself posed it in 1900 at the Paris Congress in a somewhat different way). In 1958, at the Edinburgh Congress, Nagata gave an example of a G for which $k[V]^G$ has no finite system of generators. Actually, it is natural to study the more general

---

1980 *Mathematics Subject Classification* (1985 Revision). Primary 15A72, 14L30.

situation of a rational action of G on an arbitrary affine (i.e. finitely generated and without nilpotent elements) k-algebra A (commutative, associative, and with a unity), or, geometrically, to study instead of the linear action $G:V$ the algebraic action of G on the affine variety X with coordinate algebra $k[X] = A$; see [1], [2]. Nagata considered the finite generation problem in this broad sense, calling it the generalized Hilbert's 14th problem; see [1]. It has now been completely solved:

THEOREM 1. *The following properties of a group G are equivalent*:
  a) *For all affine algebras A on which G acts rationally the algebra A^G is also affine.*
  b) *G is reductive (i.e. G has a trivial unipotent radical).*

The idea of the proof that b) $\Rightarrow$ a) when char $k = 0$ is due to Hilbert, but when char $k > 0$ is much more subtle and follows from results of Haboush [29] and Nagata [28]; see also [2]. The proof that a) $\Rightarrow$ b) was obtained by the author [3].

In contrast to the generalized problem, the original Hilbert's 14th problem is still unsolved. An important feature is that the action $G:V$ can be extended to an action of a reductive group containing G (namely $GL(V)$). This is used as follows. It can be shown that for any linear algebraic group R, closed subgroup H, and algebra A with rational action R the algebras $(A \otimes k[R/H])^R$ and A^H are isomorphic [4]. Therefore, if R is reductive, then the affineness of A^H for all such affine A is equivalent to the affineness of $k[R/H]$. Moreover, without changing A^H we may assume that H is a so-called observable group (this is equivalent to the quasiaffineness of the variety R/H [5]). We call H a Grosshans subgroup of R if H is observable and the algebra $k[R/H]$ is affine (the role of such subgroups was first noted by Grosshans in [5]). At the present time, the original Hilbert's 14th problem should be regarded as the problem of classifying the Grosshans subgroups of $GL(V)$. We should mention that the observable subgroups of reductive groups in the case char $k = 0$ were classified in 1977 by Sukhanov.

In view of what has been said, in part b) of the classical scheme for describing $k[V]^G$ it is natural to assume G is reductive. Suppose char $k = 0$. In 1868–1870 the problem was solved by Gordan for invariants of binary forms, and in 1893 by Hilbert for invariants of forms in any number of variables; these results are the high points of the classical theory. The problem of describing $k[V]^G_d$, the homogeneous component of degree d in $k[V]^G$, can be solved constructively by the methods of representation theory, and this reduces the general solution of the problem to explicitly finding a number $M_{G,V}$ such that $k[V]^G$ is generated by invariants of degree at most $M_{G,V}$; see [6]. Such an $M_{G,V}$ is now known:

THEOREM 2. *Suppose $n = \dim V$, $s = \dim G$, $r = \operatorname{rk} G$, T is a maximal torus in G, and $(k^*)^r \to T$ is a fixed isomorphism with coordinate* form $(t_1, \ldots, t_2) \mapsto (f_{ij}(t_1, \ldots, t_r))$, where $f_{ij} \in \ell(t_1, \ldots, t_r)$. Let t be the maximum of the numbers $|m_l|$, taken over all l, $1 \le l \le r$, and all monomials $t_1^{m_1} \cdots t_r^{m_r}$ occurring in at least one of the f_{ij}. For any number $h > 0$ let $C(h) = \text{LCM}\{a \in \mathbb{N} | 0 < a \le h\}$. Then we can take

1) $M_{G,V} = |G|$ if G is finite,
2) $M_{G,V} = nC(n \cdot s! \cdot t^s)$ if $G = T$ is a torus,
3) $M_{G,V} = n \cdot C(2^{r+s} n^{s+1}(n-1)^{s-r} t(s+1)!/(3^s(((s-r)/2)!)^2))$, if G is connected and semisimple.

The proof of 1) was given by Noether [7], that of 2) by Kempf [8], and that of 3) by the author [6]. The case of an arbitrary reductive group G can easily be reduced to these three [8].

The estimates of Theorem 2 are large and there remains the question of improving them. However, in any event the general solution of the problem will have only a theoretical value, since it is now clear (see below) that a practical determination of generators of a "general" algebra $\ell[V]^G$ is hardly reasonable. Nevertheless, some general results on actions distinguish whole classes of actions $G : V$ that can be thoroughly investigated (right up to an explicit description of $\ell[V]^G$). There soon arose the point of view that the main focus for invariant theory should be the classes of actions $G : V$ distinguished by various "nice" properties, whereas the other $G : V$ are "wild". We can identify three sources of such results: algebraic geometry, commutative algebra, and representation theory.

2. Geometric results on actions and the description of invariants

Suppose char $\ell = 0$ and G is a reductive group acting on an affine variety X. All of the actions considered are effective.

There long ago arose the idea of simplifying the problem of describing $\ell[X]^G$ by finding a subgroup $H \subset G$ and an H-invariant subvariety Y of X such that the restriction of functions to Y defines an isomorphism $\ell[X]^G \to \ell[Y]^H$. We may always assume that $H = N_G(Y) = \{g \in G | gY \subset Y\}$, since $\ell[Y]^H = \ell[Y]^{N_G(Y)/Z_G(Y)}$, where $Z_G(Y) = \{g \in G | gy = y \; \forall y \in Y\}$. We will call such Y Chevalley sections (apparently such Y first appeared in the classical Chevalley restriction theorem [9]).

A systematic method for constructing Chevalley sections was found by Luna and Richardson [9]. Suppose $X/G = \text{spec} \, \ell[X]^G$ and $\pi_{X,G} : X \to X/G$ is the morphism induced by the embedding $\ell[X]^G \to \ell[X]$; this morphism is G-equivariant and surjective, and each fiber $\pi_{G,X}^{-1}(\xi)$, $\xi \in X/G$, contains a unique closed orbit $T(\xi)$ [2], [10].

THEOREM 3. *Suppose X is irreducible and smooth. Then:*

a) *There exists a nonempty open set $\Omega \subset X/G$ such that $\pi_{X,G}^{-1}(\xi)$ and $\pi_{X,G}^{-1}(\eta)$ (hence also $T(\xi)$ and $T(\eta)$) are G-isomorphic $\forall \xi, \eta \in \Omega$. In* particular, there exists a reductive subgroup $C \subset G$ such that for any points $\xi \in \Omega$ and $v \in T(\xi)$ the stabilizer G_v of v is conjugate to C [10].

b) *One of the irreducible components Y of the variety $X^C = \{x \in X | Cx = x\}$ is a Chevalley section* [9].

All of this is applicable to a linear action $G : X$, for which $Y = V^C$ is a linear subspace and $N_G(Y) = N_G(C)$; Y and $N_G(Y)/Z_G(Y)$ are analogues of a Cartan subspace and the Weyl group for the general case (they become precisely these objects for the adjoint action of $G : V$). The reduction of $G : V$ to $N_G(C) : Y$ is nontautological precisely when $C \neq \{e\}$. This explains the special interest in describing those $G : V$ for which $C \neq \{e\}$. This description is based on the fact that under the conditions of Theorem 3 there exists a "stabilizer of general position", i.e. a subgroup $G_* \subset G$ such that G_v is conjugate to G_* for any v in a nonempty open subset of X; moreover, $G_* \subset C$ [30], [10]. It is simpler to describe G_* than C; on the other hand, $G_* \neq \{e\}$ implies $C \neq \{e\}$. It is clear that if G is finite or a torus, then $G_* = \{e\}$. We now know all $G : V$ with $G_* \neq \{e\}$ when either 1) G is connected and simple, or 2) G is connected and semisimple and $G : V$ is irreducible: in 1972 Èlashvili described such $G : V$ with $\dim G_* > 0$ and described $\operatorname{Lie} G_*$, and in 1975–1978 A. M. Popov considered the cases where $G_* \neq \{e\}$ is finite and described such G_* (these papers developed the ideas of Andreev, Vinberg, and Èlashvili from 1967 and those of Andreev and the author [11]). Under condition 1) or 2), for a given G there can be only a finite number of actions $G : V$ with $G_* \neq \{e\}$; without 1) or 2), this is not so, even though in the "general" case $G_* = \{e\}$ (an early result of this kind was obtained in 1967 by Khadzhiev).

Generally speaking, $G_* \neq C$. If $G_* = C$, then $G : V$ is called stable [12] (stability means that an orbit of general position is closed). A criterion for the stability of a linear action $G : V$ was obtained by the author in [12]: If G is connected and semisimple (the general case can be reduced to this one [13]), then $G : V$ is stable if and only if G_* is reductive. This means that stable actions are "typical" and enables us, under condition 1) or 2), to explicitly list all unstable actions—there are not many.

Especially close to the model situation of the adjoint action are those $G : V$ for which the "Weyl group" $N_G(Y)/Z_G(Y)$ is finite. In the known such cases, the found *ad hoc* algebra $k[V]^G$ turns out to be free. Panyushev [14] showed that these facts have a certain generality:

THEOREM 4. *Suppose G is connected and for any subvariety $Y \subset V/G$ with $\operatorname{codim} Y > 1$ we have $\operatorname{codim} \pi_{G,V}^{-1}(Y) > 1$ (the second condition is automatically satisfied if either G is semisimple, or G is contained in the orthogonal group $O(V)$, or $\pi_{G,V}$ is an equidimensional morphism). If $k[V]^G$ is isomorphic to the algebra of invariants of some finite linear group W, then both these algebras are free.*

This theorem means that neither the theory of invariants of connected linear groups nor the theory of invariants of finite linear groups can be reduced to the other, and confirms the view of recent years that these are two "different" theories.

There exist examples of proper linear subspaces of V that are Chevalley sections, but cannot be obtained by applying the construction of Theorem 3 ($C = \{e\}$). Apart from Theorem 3, there are no general results on constructing Chevalley sections. Fundamental to the proof of Theorem 3 and many other results on the actions $G : X$ is Luna's slice theorem [10].

3. Commutative algebra and invariant theory

We can now distinguish two approaches to the more subtle (than finding generators) problem of describing the multiplicative structure of the algebra of invariants $\mathscr{k}[V]^G$. The first was inspired by a result of Hochster and Roberts [15]:

THEOREM 5. *If X is smooth, then $\mathscr{k}[X]^G$ is a Cohen-Macaulay algebra.*

In fact, as was shown by Boutot [16], we have the more general

THEOREM 6. *If X has rational singularities, then so does X/G.*

For a linear action $G : V$, Theorem 5 implies the existence of homogeneous elements $a_1, \ldots, a_m, b_1, \ldots, b_s \in \mathscr{k}[V]^G$ such that $a_1, \ldots, a_m$ (the parameters) are algebraically independent and

$$\mathscr{k}[V]^G = \bigoplus_{i=1}^{s} \mathscr{k}[a_1, \ldots, a_m]b_i.$$

This reduces the description of the structure of $\mathscr{k}[V]^G$ to a definition of the "multiplication table" for the b_i and provides a formula for the Poincaré series $P_{G,V}(t) = \sum_{d \geq 0}(\dim \mathscr{k}[V]^G_d)t^d = \left(\sum_{i=1}^{s} t^{\deg b_i}\right) / \left(\prod_{j=1}^{m}(1 - t^{\deg a_j})\right)$. Since $\deg P_{G,V}(t) \leq 0$, as Kempf showed in 1979, this means that $\max_i b_i \leq \sum_{j=1}^{m} \deg a_j$. Consequently, $b_1, \ldots, b_s$ can be constructively found (in theory) for known $a_1, \ldots, a_m$. The constructive determination of $a_1, \ldots, a_m$ is a complicated problem. For a finite G it was solved by Dade; see [24] (one chooses $f_i \in V^*$, $1 \leq i \leq n$, so that $f_1 \neq 0$ and $f_{i+1} \notin \langle g_1 f_1, \ldots, g_i f_i \rangle$ $\forall g_j \in G$; then a_i is the product of all elements of the orbit Gf_i). For a connected G there is no known recipe of this kind, but there is an idea, going back to Hilbert [31], based on the characterization of parameters as algebraically independent, homogeneous invariants for which $\{v \in V | a_i(v) = 0 \; \forall i\}$ is equal to the "null-cone" $\pi_{G,V}^{-1}(\pi_{G,V}(0))$ and on the geometric description of $\pi_{G,V}^{-1}(\pi_{G,V}(0))$ afforded by the Hilbert-Mumford theorem [2]. By this method we obtain explicit upper bounds on $\deg a_i$ and a proof of parts (2) and (3) of Theorem 2; see [8], [6].

The second approach is provided by the classical theory of syzygies. Suppose $y_1, \ldots, y_p$ is a minimal system of homogeneous generators of $k[V]^G$ (the number p is the embedding dimension $\operatorname{ed} k[V]^G$ of the algebra $k[V]^G$), $A = k[t_1, \ldots, t_p]$ is the polynomial algebra, graded by the condition $\deg t_i = \deg y_i$, and I is the kernel of the epimorphism $A \to k[V]^G$, $t_i \mapsto y_i$. This I is the first syzygy module of $k[V]^G$; its elements (syzygies) are relations among $y_1, \ldots, y_p$. By choosing a minimal homogeneous system of generators of the A-module I we can analogously represent I as the image of a free A-module M_1 under a homogeneous epimorphism $\varphi_1: M_1 \to I$, then treat $\operatorname{Ker} \varphi_1$ in the same way, and so on. We obtain as a result a free minimal resolution of the A-module I, which is finite by the Hilbert syzygy theorem:

$$0 \to M_h \xrightarrow{\varphi_h} \cdots \to M_1 \xrightarrow{\varphi_1} I \xrightarrow{\varphi_0} 0.$$

Its length h is the homological dimension $\operatorname{hd} k[V]^G$ of the algebra $k[V]^G$, and $\operatorname{Ker} \varphi_{i-1}$ is the ith syzygy module. By Theorem 5, $h = p - \dim V/G$.

A complete description of all $\operatorname{Ker} \varphi_i$ has been obtained only in a few cases. In the 19th century it was obtained only for invariants of binary forms of degree $d \leq 6$ (and in some other elementary cases), and the case $d = 8$ was successfully analyzed only as recently as 1967 by Shioda. The older works described the actions of the classical groups on systems of vectors and covectors, but the higher syzygies became known only after the papers of Lascoux in 1978 and the subsequent papers of others. The reason, of course, is the complexity of the structure of $k[V]^G$ in the "general case". This was undoubtedly understood long ago, but the precise meaning of the assertion has become clear only recently.

A natural measure of the complexity of $k[V]^G$ is the number $h = \operatorname{hd} k[V]^G$. It can be taken as the foundation of the systematic investigation of the algebras $k[V]^G$ for various actions $G: V$ of a given group G. Instead of a random choice of $G: V$ one should first consider those actions for which h is small. Such a program was firmly established by the following theorem, proved by the author in [12], [23].

THEOREM 7. *Suppose G is either connected and semisimple or finite. Then for any integer $d \geq 0$ there exist, to within the isomorphism and addition of a trivial direct summand, only finitely many finite-dimensional rational G-modules V such that $\operatorname{hd} k[V]^G = d$.*

It follows that if $G \neq \{e\}$, then there exists a G-module V such that $\operatorname{hd} k[V]^G > d$ (and there are infinitely many such V). This gives a precise meaning to the assertion on the complexity of $k[V]^G$ in the "general case". The estimates in [13] also provide it with a quantitative expression. For example, if $G = SL_2$ and $V = S_n$ is the module of binary forms of degree

n, then for any prime p with $n > 4p + 5$ we have

$$\text{hd}\,k[S_n]^{SL_2} \geq (p-1)\binom{[n/p]+p-2}{p} + \frac{p-1}{2}([n/p]-1)^2[n/p] - n + 2,$$

so that $\text{hd}\,k[S_n]^{SL_2}$ grows faster than any power of n as $n \to \infty$. If G is an exceptional simple group and V is different from the adjoint module, the elementary one or its dual, then $\text{hd}\,k[V]^G \geq 9208, 26335, 3403, 2278$, or 3 for $G = E_8, E_7, E_6, F_4$, or G_2, respectively. Moreover, the assertion of Theorem 7 and the constructive nature of its proof show that the situation is actually not so hopeless as it could be a priori and make it natural to pose the problem of classifying all $G : V$ with a prescribed value of $d = \text{hd}\,k[V]^G$. Such a classification should be carried out by successively increasing d, since the method itself presupposes the possibility of induction: an important role in this method is played by the monotonicity theorems, according to which, in particular, ed and hd do not increase on passing to subrepresentations and slice-representations. They are a manifestation of a general "inheritance principle", according to which the properties of $k[V]^G$ do not worsen (are inherited) under the indicated passages; there is now much evidence of its usefulness.

The first step in this program is to describe those $G : V$ with $\text{hd}\,k[V]^G = 0$, i.e. with a free algebra of invariants (property (FI)). If G is finite, the answer is given by

THEOREM 8. *Suppose G is a finite group, $G \subset GL(V)$. Then the following properties are equivalent*:
 a) *G is generated by pseudoreflections*;
 b) *$k[V]^G$ is free*.

The proof was given in 1954 by Shephard and Todd, who also explicitly classified (by a list) all G with property a), and in 1955 by Chevalley, who proved a) $\Rightarrow$ b).

For connected G a method for classifying those $G : V$ with a free algebra $k[V]^G$ was found only rather recently by Vinberg, Kac, and the author [25]; it was tested in that paper by the example of connected simple G and irreducible $G : V$ (the method is based on the "inheritance principle" of property (FI)).

THEOREM 9. *Suppose G is a connected simple irreducible linear group. Then the following properties are equivalent*:
 1) *the algebra of invariants of G is free*;
 2) *G is one of the following groups*;
 a) *the case $k[V]^G = k$*: SL_n, Sp_n (n even), $\Lambda^2 SL_n$ (n odd), Spin_{10};
 b) *the case* $\dim V/G = 1$: SO_n ($n \neq 4$), $\Lambda^2 SL_n$ (n even), $S^2 SL_n$, $\Lambda^3 SL_n$ ($n = 6, 7, 8$), $\Lambda^3_0 Sp_6$, $S^3 SL_2$, Spin_n ($n = 7, 9, 11, 12, 14$), E_6, E_7, G_2;

c) *adjoint groups of simple Lie groups*;

d) *isotropy groups of symmetric spaces*: $\Lambda_0^2 Sp_n$ (n even), $S_0^2 SO_n$, $\Lambda^4 SL_8$, $\Lambda_0^4 Sp_8$, $Spin_{16}$, F_4;

e) *the other groups*: $\Lambda^3 SL_9$, $S^3 SL_3$, $Spin_{13}$ (here $\Lambda_0^m G$ and $S_0^m G$ denote the highest irreducible constituent of the linear groups $\Lambda^m G$ and $S^m G$, respectively).

By the same method, Schwarz in 1978 and, independently, Adamovich and Golovina in 1979 classified the reducible actions $G : V$ of connected simple G with a free $\mathscr{k}[V]^G$.

Whether it is possible to describe those $G : V$ with a free $\mathscr{k}[V]^G$ and connected G not by a list, but by a single group-theoretic characterization (as in Theorem 8), it is not known (a naive attempt to extend the implication a) $\Rightarrow$ b) of Theorem 8 to infinite G is, as Zalesskii showed in 1983, untenable). However, we do know several general constructions of $G : V$ with a free $\mathscr{k}[V]^G$. The first is that of the θ-groups introduced by Vinberg [26]. They are constructed from a semisimple automorphism θ of a connected reductive group G as the linear groups defined by the adjoint action of $(G^\theta)^0$ on proper subspaces of $d\theta$ in $\operatorname{Lie} G$. When $\theta = e$ we obtain the adjoint groups, and when $\theta^2 = e$, $\theta \neq \operatorname{Id}$, the isotropy groups of symmetric spaces, so that Chevalley's classical results on such groups are consequences of the theory [26]. These groups belong to a wider class of groups with a free $\mathscr{k}[V]^G$: the class of polar groups of Dadok and Kac [27] (these groups are defined by the condition that there exist a closed orbit Gv such that $\dim\{x \in V | (\operatorname{Lie} G)x \subseteq (\operatorname{Lie} G)v\} = \dim V/G$). The polarity property (Pol) is hereditary, and in [27] there can be found, in addition to the general theory, a list of connected simple linear G with property (Pol). All groups in the list of Theorem 9 are polar, and almost all are related to θ-groups. The third construction is due to Panyushev: the isotropy groups of homogeneous spaces G/H, where $G \supset H$ are connected reductive groups and $G \supset H$ is a spherical pair (i.e. a Borel subgroup B of G has a dense open orbit in G/H): in 1985 Mikityuk and Brion, independently and by different methods, classified such pairs. Polar groups have a (linear) Chevalley section with finite "Weyl group", and the freeness of $\mathscr{k}[V]^G$ follows from Theorem 4.

The next step is $\operatorname{hd} \mathscr{k}[V]^G = 1$, i.e. V/G is a hypersurface. Here (as in the case $\operatorname{hd} \mathscr{k}[V]^G = 2$, if G is either connected and semisimple, or finite and lies in $SL(V)$) $\mathscr{k}[V]^G$ is a complete intersection (i.e. $\operatorname{rk} M_1 = \operatorname{ed} \mathscr{k}[V]^G - \dim V/G$). This property is hereditary. In 1982 Kac and Watanabe and, independently, Gordeev established an analogue of the implication b) $\Rightarrow$ a) of Theorem 8

THEOREM 10. *If G is a finite subgroup of $GL(V)$ and $\mathscr{k}[V]^G$ is a complete intersection, then G is generated by the set* $\{g \in G | \operatorname{rank}(g - 1) \leq 2\}$.

The idea of the proof is similar to that of Theorem 4 and is based on the fact that a certain open subset of V/G is simply connected. (It is worth noting here that, as Mikhaĭlova and Shvartsman proved in 1985, a finite group $H \subset GL(\mathbb{R}^n)$ is generated by the set $\{g \in H | \operatorname{rank}(g - 1) = 2\}$ precisely when $\mathbb{R}^n/H$ is homeomorphic to $\mathbb{R}^n$.) These facts enable us to obtain some important information. We now have a complete classification (list) of the finite linear groups $G \subset GL(V)$ for which $k[V]^G$ is a complete intersection (Gordeev and Nakajima, 1985–1986) and, in particular, a hypersurface (Nakajima, 1983). In 1984 Nakajima found all connected simple irreducible linear G with the same property; such groups of type $S^d SL_n$ had been found earlier by Beklemishev and, independently, by Kac in 1983.

At present, the cases of larger $\operatorname{hd} k[V]^G$ have been investigated only in the classical situation of invariants of binary forms; in [17] the author proved

THEOREM 11. 1) $\operatorname{hd} k[S_n]^{SL_2} \leq 10$ *precisely for* $n = 0, 1, 2, 3, 4, 5, 6, 8$ (*and* $\operatorname{hd} k[S_n]^{SL_2} \geq 18, 34, 28$ *for* $n = 7, 10, 12$, *respectively*).

2) *When* $V^{SL_2} = 0$, *the condition* $\operatorname{hd} k[V]^{SL_2} = h \leq 3$ *holds precisely in the following cases:*
 a) $h = 0 : V = S_1, S_2, S_3, S_4, 2S_1, S_1 + S_2, 2S_2, 3S_1$;
 b) $h = 1 : V = S_5, S_6, S_2 + S_3, S_2 + S_4, 2S_1 + S_2, S_1 + 2S_2, 3S_2, S_1 + S_3, S_1 + S_4, 2S_4, 4S_1$;
 c) $h = 2 : V = 2S_3$;
 d) $h = 3 : V = S_8, 5S_1$.

This means that the classics of the 19th century, having essentially investigated those cases that admitted investigation, did not omit any that were sufficiently simple.

Regarding the hierarchy of "nice" properties of algebras: (FA), a free algebra $\Rightarrow$ (H), a hypersurface $\Rightarrow$ (CI), a complete intersection $\Rightarrow$ (Gor), a Gorenstein algebra $\Rightarrow$ (CM), a Cohen-Macaulay algebra, we now have a satisfactory description of the $k[V]^G$ with these properties (the property (Gor) holds if G is either connected and semisimple [15], or finite and lies in $SL(V)$, and for finite G containing no pseudoreflections, $k[V]^G$ is Gorenstein implies $G \subset SL(V)$; see [24]). According to Stanley (1978), $k[V]^G$ being Gorenstein is equivalent to a functional equation for the Poincaré series:

$$P_{G,V}(t^{-1}) = (-1)^{\dim V/G} t^{-\deg P_{G,V}} P_{G,V}(t).$$

4. Representation theory and invariant theory

An invariant $f \in k[X]^G$ of an action $G : X$ can be viewed as an equivariant morphism $f : X \to k$. Replacing k by any G-module W, we obtain the more general concept of covariant. The presence of a covariant is equivalent to the existence of a G-module homomorphism $W^* \to k[X]$.

Therefore the question of the structure of covariants reduces to that of describing the G-module structure of $\mathscr{k}[X]$. It is convenient to regard $\mathscr{k}[X]$ as a $\mathscr{k}[X]^G$-module (the module of covariants): if $\mathfrak{E}$ is the set of classes of equivalent simple rational G-modules and $\mathscr{k}[X]_\lambda$ is an isotypical component of $\mathscr{k}[X]$ of type $\lambda \in \mathfrak{E}$, then $\mathscr{k}[X]_\lambda$ is a $\mathscr{k}[X]^G$-module of finite type and $\mathscr{k}[X] = \bigoplus_{\lambda \in \mathfrak{E}} \mathscr{k}[X]_\lambda$. If G is connected, then $\mathfrak{E}$ can be identified with the monoid of dominant weights relative to B, and the highest vectors in $\mathscr{k}[X]$ form the so-called subalgebra of covariants $\mathscr{k}[X]^u$ (U is the unipotent radical of B). In the classical situation of a linear action $G : V$, the case of a free module of covariants (property (FM)) is of particular interest. In this case $\mathscr{k}[V] = H \oplus \mathscr{k}[V]^G$, where H is an invariant complement in $\mathscr{k}[V]$ of the ideal generated by $\bigoplus_{d \geq 1} \mathscr{k}[V]_d^G$, and $\mathscr{k}[V]_\lambda$ is a free $\mathscr{k}[V]^G$-module of rank m_λ equal to the multiplicity of the occurrence of a G-module W of type λ in H. As Schwarz showed in 1978, m_λ can be explicitly found from C (see Theorem 3): V is the sum of a trivial C-module, the isotropy C-module G/C, and a C-module L with $L^C = 0$, and $m_\lambda = \dim(\mathscr{k}[L] \otimes W^*)^C$. In the case $C = G_*$ the answer is particularly simple: $L = 0$ and $m_\lambda = \dim(W^*)^C$.

The problem of describing those $G : V$ with property (FM) was first considered by Kostant [18], who investigated in detail the adjoint representation. A general study was begun by the author in [19]; it was based on the following criterion:

THEOREM 12. *Property* (FM) *holds for* $G : V$ *if and only if* $\mathscr{k}[V]^G$ *is a free algebra and* $\pi_{G,V}$ *is an equidimensional morphism.*

Thus for a finite G, property (FM) holds if and only if G is generated by pseudoreflections; in this situation, $G : H$ is equivalent to the regular action. If G is connected, equidimensionality of $\pi_{G,V}$ is an independent geometric condition that does not follow from the freeness of $\mathscr{k}[V]^G$. This naturally leads to the independent problem of classifying all $G : V$ with an equidimensional $\pi_{G,V}$ (property (E)). The "null-cone" $\pi_{G,V}^{-1}(\pi_{G,V}(0))$ is a fiber of maximal dimension, and from its geometric description we can obtain lower bounds for $\dim(\pi_{G,V}^{-1}(\pi_{G,V}(0))$; see [19]. Moreover, (FM) and (E) are hereditary properties. In this direction we have

THEOREM 13. *Suppose* G *is either connected and semisimple or finite. Then, to within isomorphism and addition of a trivial direct summand, there exist only finitely many G-modules* V *with property* (E) *(and therefore with property* (FM)).

This assertion admits a quantitative refinement; for example, if G is connected and semisimple and $V^G = 0$, then $\dim V \leq 3 \dim G - 3$. Indeed in many cases it is possible to give a complete classification: in [19] the author listed all irreducible $G : V$ with a connected simple G and with property (E) or (FM); such lists for reducible $G : V$ with a connected simple G were then found by Schwarz in 1978 and Adamovich in 1980, and for an irreducible $G : V$ with a connected semisimple G by Littelmann in 1986. A general group-theoretic characterization (not a list) of $G : V$ with property (E) or (FM) is not known, but there are several general constructions: θ-groups, polar groups, and isotropy groups of reductive spherical pairs have properties (E) and (FM). Property (E) holds a priori for the so-called observable actions (property (FO)), i.e. those for which each fiber of $\pi_{G,V}$ contains only a finite number of orbits; the irreducible $G : V$ with a connected G and property (FO) were found in 1975 by Kac. On the model of the adjoint action we can consider two more "nice" properties:

1) (SC): some linear subspace of V is a Chevalley section with finite "Weyl group";

2) (SW): V contains a linear variety ("Weierstrass section") that intersects each fiber of $\pi_{G,V}$ in a single point.

A priori, (SC) $\Rightarrow$ (FM) $\Leftarrow$ (SW). A comparison of lists reveals the following remarkable coincidence:

THEOREM 14. *Suppose G is connected and simple and $G : V$ is irreducible. Then the following properties are equivalent*: 1) (FI); 2) (FM); 3) (E); 4) (FO); 5) (SC); 6) (SW); 7) $G_* = \{e\}$; 8) (Pol); 9) *G is a group in the list of Theorem* 9.

This result does not carry over to the general case (G is connected and semisimple and $G : V$ is not necessarily irreducible), but there are various implications among the indicated properties. In [19] the author stated a general conjecture (now sometimes called by specialists the "Russian conjecture"): If G is connected, then (E) $\Leftrightarrow$ (FM). A comparison of classification results supports it when G is simple or when G is semisimple and $G : V$ is irreducible; obtaining a proof is an intriguing and, apparently, complex problem.

Apart from these, there is at least one other "nice" property (FC): freeness of the algebra of covariants $k[V]^U$. A priori, (FC) $\Rightarrow$ (FM). All cases of $G : V$ with a connected simple G and property (FC) were listed in 1985 by Brion. Also in 1985 Panyushev found for the algebra of covariants an analogue of the construction of Theorem 3 resulting in a "Chevalley section". For any action $G : X$ the algebras $k[X]$ and $k[X]^U$ have a number of important common properties (e.g. finite generation, integral closure, rationality of singularities) [4]. From $k[X]^U$ we can recover $k[X]$ to some extent, which enables us to represent any action as a plane deformation (with a one-dimensional basis) of some very special action [4], [20]. This, in turn, reduces questions on $G : X$ to special actions [20].

The fact that only a few $G : V$ possess some "nice" property leads us to hope that we can effectively solve for these actions the main problems of invariant theory: a classification of orbits and a description of the algebras $k[V]^G$. This is indeed happening: apart from the classic cases, solutions have been found in our time for Spin_n, $n \leq 12$ (Igusa, 1970), Spin_{13} (Gatti and Viniberghi [Vinberg], 1978), Spin_{14} (the author, 1978), Spin_{16} (Antonyan and Èlashvili, 1982), $\Lambda^3 SL_q$ (Vinberg and Èlashvili, 1978), $\Lambda^4 SL_8$ (Antonyan, 1980), and $S^3 SL_4$ (Beklemishev, 1982; see also the papers of Igusa (1973) on the geometry of so-called "absolutely admissible" representations).

5. Birational invariant theory

The main problem here is whether the field $k(V)^G$ of invariant rational functions on V is rational. In essence it was posed by E. Noether in 1913. As usual, k is algebraically closed and $\mathrm{char}\, k = 0$. Recently substantial progress has been made on this problem. Saltman [21] constructed an example of a finite G such that $k(V)^G$ is not retract rational and therefore not rational. He did it by considering a certain subgroup Br_0 of $Br(k(V)^G)$ which must be trivial if $k(V)$ is retract rational, but is nontrivial in his example. It turns out that Br_0 depends only on G, and not on $G:V$ (this was essentially established by Faddeev in 1951, since all $G:V$ are retract isomorphic for different $G:V$).

In 1986 Bogomolov showed that $Br_0 = \{\gamma \in H^2(G, \mathbb{Q}/\mathbb{Z}) | \gamma_A = 0\}$ for each Abelian subgroup $A \subset G$. At present we do not have any example of a nonrational $k(V)^G$ with a connected G. In 1982 Vinberg proved that $k(V)^G$ is rational when G is connected and solvable, and in 1986 Bogomolov obtained the same conclusion when G is a connected, simply connected, reductive group without simple factors of type B_n, D_n, E_8 and $G_* = \{e\}$. An important step was taken by Katsylo [22], who introduced an appropriate analogue of a Chevalley section, namely, an irreducible subvariety $Y \subset V$ such that $\overline{GY} = V$ and with the property that the conditions $g(y) \in Y$, $g \in G$, $y \in Y$ imply $g \in N_G(Y)$. Restriction of functions defines an isomorphism $k(V)^G \to k(Y)^{N_G(Y)}$; another important property is the opportunity to "multiply" such Y by covariants $W \to V$. By considering such "sections" Katsylo proved in 1984 that $k(V)^G$ is rational for almost all actions of $G = SL_2$ and $G = SL_2 \times k^*$ (k^* acts by dilations on V) and, as a consequence, that the variety of moduli of hyperelliptic curves of genus g is rational (for $g = 4$ some additional considerations are needed).

It is worth noting that the problem is closely related to that of the rationality of homogeneous spaces: $k(V)^G$ is retract rational for $G:V$ with $G_* = \{e\}$ if and only if GL_n/G is retract rational for some embedding $G \subset GL_n$ (so that, by Saltman's example, there exist nonrational G/H with a connected G and finite H; whether this is possible with a connected reductive H is not known).

Bibliography

1. M. Nagata, *Lectures on the fourteenth problem of Hilbert*, Tata Institute of Fundamental Research, Bombay, 1965.
2. D. Mumford and J. Fogarty, *Geometric invariant theory*, Ergeb. Math. Grenzgeb., Vol. 34, Springer-Verlag, Berlin, 1982.
3. V. L. Popov, *Hilbert's theorem on invariants*, Dokl. Akad. Nauk SSSR **249** (1979), No. 3, 551-555; English transl. in Soviet Math. Dokl. **20** (1979), 1313-1321.
4. ___, *Contraction of the actions of reductive algebraic groups*, Mat. Sb. **130** (1986), No. 3, 310-334; English transl. in Math. USSR-Sb. **58** (1987), 311-335.
5. F. Grosshans, *Observable groups and Hilbert's fourteenth problem*, Amer. J. Math. **95** (1973), No. 1, 229-253.
6. V. L. Popov, *The constructive theory of invariants*, Izv. Akad. Nauk SSSR Ser. Mat. **45** (1981), 1100-1120; English transl. in Math. USSR-Izv. **19** (1982), 359-376.
7. E. Noether, *Der Endlichkeitssatz der Invarianten endlicher Gruppen*, Math. Ann. **77** (1915), No. 1, 89-92.
8. G. R. Kempf, *Computing invariants*, Preprint, Johns Hopkins Univ., 1986.
9. D. Luna and R. W. Richardson, *A generalization of the Chevalley restriction theorem*, Duke Math. J. **46** (1979), No. 3, 487-496.
10. D. Luna, *Slices étales*, Bull. Soc. Math. France Mém. 33 (1973), 81-105.
11. E. M. Andreev and V. L. Popov, *On stationary subgroups of points in general position in the representation space of a semisimple Lie group*, Funktsional. Anal. i Prilozhen., **5** (1971), No. 4, 1-8; English transl. in Functional Anal. Appl. **5** (1971).
12. V. L. Popov, *Stability criteria for the action of a semisimple group on a factorial manifold*, Izv. Akad. Nauk SSSR Ser. Mat. **34** (1970), 523-531; English transl. in Math. USSR-Izv. **4** (1970), 527-535.
13. ___, *Syzygies in the theory of invariants*, Izv. Akad. Nauk SSSR Ser. Mat. **47** (1983), 544-622; English transl. in Math. USSR-Izv. **22** (1984), 507-585.
14. D. I. Panyushev, *On orbit spaces of finite and connected linear groups*, Izv. Akad. Nauk SSSR Ser. Mat. **46** (1982), 95-99; English transl. in Math. USSR-Izv. **20** (1983), 97-101.
15. M. Hochster and J. Roberts, *Rings of invariants of reductive groups acting on regular rings are Cohen-Macaulay*, Adv. Math. **13** (1974), 125-175.
16. J.-F. Boutot, *Singularités rationneles et quotients par les groupes réductifs*, Preprint, Strasbourg Univ., 1982.
17. V. L. Popov, *Homological dimension of algebras of invariants*, J. Reine Angew. Math. **341** (1983), 157-173.
18. B. Kostant, *Lie group representations on polynomial rings*, Amer. J. Math. **85** (1963), 327-404.
19. V. L. Popov, *Representations with a free module of covariants*, Funktsional. Anal. i Prilozhen. **10** (1976), No. 3, 91-92; English transl. in Functional Anal. Appl. **10** (1976).
20. È. B. Vinberg, *Complexity of actions of reductive groups*, Funktsional. Anal. i Prilozhen. **20** (1986), No. 1, 1-13; English transl. in Functional Anal. Appl. **20** (1986).
21. D. J. Saltman, *Noether's problem over an algebraically closed field*, Invent. Math. **77** (1984), No. 1, 71-84.
22. P. I. Katsylo, *On rationality of moduli spaces of hyperelliptic curves*, Izv. Akad. Nauk SSSR Ser. Mat. **48** (1984), English transl. in Math. USSR-Izv. **25** (1985), 45-50.
23. V. L. Popov, *A finiteness theorem for representations with a free algebra of invariants*, Izv. Akad. Nauk SSSR Ser. Mat. **46** (1982), 347-370; English transl. in Math. USSR-Izv. **20** (1983), 333-354.
24. R. P. Stanley, *Invariants of finite groups and their applications to combinatorics*, Bull. Amer. Math. Soc. **1** (1979), No. 3, 475-511.
25. V. G. Kac, V. L. Popov, and E. B. Vinberg, *Sur les groupes algébriques dont l'algèbre des invariants est libre*, C. R. Acad. Sci. Paris Sér A, **283** (1976), 875-878.

26. È. B.Vinberg, *The Weyl group of a graded Lie algebra*, Izv. Akad. Nauk SSSR Ser. Mat. **40** (1976), 488–526; English transl. in Math. USSR-Izv. **10** (1976), 463–495.

27. J. Dadok and V. Kac, *Polar representations*, J. Algebra **92** (1985), No. 2, 504–524.

28. M. Nagata, *Invariants of a group in an affine ring*, J. Math. Kyoto Univ. **3** (1964), 369–377.

29. W. Haboush, *Reductive groups are geometrically reductive*, Ann. Math. **102** (1975), 67–83.

30. R. W. Richardson, *Principal orbit types for algebraic transformation groups in characteristic zero*, Invent. Math. **16** (1972), 6–14.

31. D. Hilbert, *Über die vollen Invariantensysteme*, Math. Ann. **42** (1893), 313–373.

Translated by G. A. KANDALL